# DIGITAL SIGNAL PROCESSING LABORATORY

# DIGITAL SIGNAL PROCESSING LABORATORY


B. Preetham Kumar

California State University
Department of Electrical
and Electronic Engineering
Sacramento, CA

A CRC title, part of the Taylor & Francis imprint, a member of the Taylor & Francis Group, the academic division of T&F Informa plc.

**Library of Congress Cataloging-in-Publication Data**

Kumar, B. Preetham.
Digital signal processing laboratory / B. Preetham Kumar.
p. cm.
Includes bibliographical references and index.
ISBN 0-8493-2784-9 (alk. paper)
1. Signal processing—Digital techniques—Textbooks. I. Title.

TK5102.9.K835 2005
621.382′2—dc22 2004058495



Direct all inquiries to CRC Press, 2000 N.W. Corporate Blvd., Boca Raton, Florida 33431.



**Visit the CRC Press Web site at www.crcpress.com**


International Standard Book Number 0-8493-2784-9
Library of Congress Card Number 2004058495
Printed in the United States of America 1 2 3 4 5 6 7 8 9 0
Printed on acid-free paper

*To Veena and Vasanth*

*and*

*In memory of my parents*

# Preface

The motivating factor in the preparation of this book was to develop a practical, and readily understandable laboratory volume in Digital Signal Processing (DSP). The intended audience is primarily undergraduate and graduate students taking DSP for the first time as an elective course. The book is very relevant at the present time, when software and hardware developments in DSP are very rapid, and it is vital for the students to complement theory with practical software and hardware applications in their curriculum.

This book evolved from study material in two courses taught at the Department of Electrical and Electronic Engineering, California State University, Sacramento (CSUS). These courses, Introduction to Digital Signal Processing and Digital Signal Processing Laboratory, have been offered at CSUS for the past several years. During these years of DSP theory and laboratory instruction for senior undergraduate and graduate students, often with varied subject backgrounds, we gained a great deal of experience and insight. Students who took these courses gave very useful feedback, such as their interest in an integrated approach to DSP teaching that would consist of side-by-side training in both theory and practical software/hardware aspects of DSP. In their opinion, the practical component of the DSP course curriculum greatly enhances the understanding of the basic theory and principles.

The above factors motivated me to prepare the chapters of this book to include the following components: a *brief theory* to explain the underlying mathematics and principles, a *problem solving* section with a reasonable number of problems to be worked by the student, a *computer laboratory* with programming examples and exercises in MATLAB® and Simulink®, and finally, in applicable chapters, a *hardware laboratory,* with exercises using test and measurement equipment and the Texas Instruments TMS320C6711 DSP Starter Kit.

In Chapter 1, we go into a brief theory of DSP applications and systems, with solved and unsolved examples, followed by a computer lab, which introduces the students to basic programming in MATLAB, and creation of system models in Simulink®. This chapter concludes with a hardware section, which contains instructions and exercises on usage of basic signal sources, such as synthesized sweep generators, and measuring equipment, such as oscilloscopes and spectrum analyzers.

Chapter 2 is a more detailed description of Linear Time Invariant (LTI) discrete-time signals and systems and the mathematical tools used to describe these systems. Basic concepts such as Z-transform, system function, discrete-time convolution, and difference equations are reviewed in the theory section. Practical types of LTI systems, such as inverse systems and minimum-phase systems are also discussed, with example problems. This is followed by a computer lab, which has guidance and exercises in the creation and simulation of LTI system models.

Chapter 3 covers practical time and frequency analysis of discrete-time signals, with emphasis on the evolution of the Discrete Fourier Transform (DFT) and the Fast Fourier Transform (FFT). The software lab includes spectral analysis, using FFT, of practical periodic and nonperiodic signals, such as noisy signal generators, and amplitude modulation (AM) systems. The hardware lab involves actual measurement of harmonic distortion in signal generators, spectrum of AM signals, and the comparison of measured results with simulation from the computer lab section.

Chapter 4 is a practical discussion of the analog-to-digital (A/D) process, with an initial brief review of sampling, quantization (uniform and nonuniform), and binary encoding in the Pulse Code Modulation (PCM) process. The software lab includes MATLAB/Simulink® A/D process simulation of practical audio signals, and advanced systems such as differential PCM. The hardware lab gives guidance of the construction and testing of a FET Sample and Hold circuit.

Chapter 5 and Chapter 6 are devoted to design and application of digital filters. Chapter 5 reviews the basic concepts of digital filters and analytical design techniques for FIR and IIR digital filter design. The computer lab details MATLAB CAD techniques for Finite Impulse Response (FIR) and Infinite Impulse Response (IIR) digital filters and has a series of rigorous exercises in usage of these techniques. Chapter 6 deals with the application of digital filters to one-dimensional (audio) and two-dimensional (video) signals. The computer lab has a set of practical exercises in the application of one- and two-dimensional digital filters for practical purposes, such as audio recovery from noise and image deblurring.

Chapter 7 and Chapter 8 are focused on the application of practical DSP processes through digital signal processor (DSP) hardware. The hardware used in this book is the Texas Instruments TMS320C6711 Digital Signal Processor Starter Kit. Chapter 7 deals in detail with the organization and usage of the 6711 DSK, with a set of practical introductory exercises, such as signal generation and filtering. Chapter 8 is more applied and covers the hardware application and programming of the 6711 DSK for practical filtering applications of noise from audio signals.

There are six appendices. The first four appendices give detailed hardware descriptions and user instructions for the equipment used in this book. The four equipment models covered are synthesized sweep generators, spectrum analyzers, dynamic signal analyzers, and digitizing oscilloscopes in Appendices A, B, C, and D, respectively. Appendix E gives detailed schematics, hardware description, and user instructions on the Texas Instruments 6711 DSK. Finally, Appendix F gives brief descriptions of alternative equipment and manufacturers who produce equipment with similar capabilities as the ones described in Appendices A–D.

I would like to thank a number of people, without whom this book would not have been completed. First of all, I greatly appreciate the help from Stan Wakefield, publishing consultant, who initiated my contact with CRC Press. I am very thankful to CRC acquisitions editor, Nora Konokpa, for her constant advice and encouragement throughout the manuscript preparation process. I would also like to thank Helena Redshaw and Jessica Vakili of CRC Press for guiding me in the preparation of the different chapters of the book. I would like to thank all the students at CSUS, who, over the years, gave very useful feedback on the DSP courses, which formed the basis of this book. I am particularly indebted to my student, Nilesh Lal, who tested and debugged all the experiments on the TI 6711 DSK, which contributed to the last, but most practical, sections of the book.

Finally, I would like to thank my wife, Priya, who took time off her already very busy schedule to proofread the chapters before submission to CRC Press. Above all, I am grateful for her help and encouragement in whatever I have attempted over the years.

# Contents

# *Note to Readers on Structure of Book and Exercises*

This book is organized into *eight* chapters and *six* appendices, with each chapter typically having the following three sections: *brief theory, computer laboratory* and *hardware laboratory.* All eight chapters have theory and computer laboratory sections; however Chapters 2, 5 and 6 do *not* have a hardware section. Generally, each chapter includes a brief theory section, followed by a MATLAB/Simulink® simulation section, and finally, the hardware section, which includes experiments on generation and measurement of signals using signal generators, digital oscilloscopes and spectrum analyzers, and the Texas Instruments' TMS320C6711 Digital Signal Processor Starter Kit.

This three-pronged approach is aimed at taking students from theory, to simulation, to experiment, in a very effective way. Additionally, instructors have the option of selecting only the computer laboratory, or hardware laboratory, or both, for their individual classes based on availability of software or hardware.

## Guidelines for Instructors

Please note that in each chapter, each of the three sections (theory, computer lab and hardware lab) have ***exercises*** for students. However, these exercises are numbered starting from the theory section and proceeding sequentially to the hardware section. Hence, each chapter typically has about four to five ***exercises*** each, and the instructor can assign any or all of the exercises for the student.

## Guidelines for Students

Please attempt ***all exercises*** systematically, or as assigned by your instructor, after reviewing the theory material in each chapter. Clarify all doubts with the instructor before proceeding to the next section since each section draws information from the previous material.

# 1

# Introduction to Digital Signal Processing

## 1.1 Brief Theory of DSP Concepts

Digital Signal Processing (DSP) primarily deals with commonly occurring signals such as speech, music, video, EKG or ECG (heart), and EEE (brain).[1] These signals occur naturally in analog continuous-time form, as shown in Figure 1.1, which depicts a typical speech signal, representing the phrase "She sells sea shells" said over a time span of about 1.8 seconds.

### 1.1.1 Applications of DSP

Given an analog signal (see Figure 1.1), the following applications are possible:

- *Preparing the analog signal for communication through a communication channel.* Years of study and research have shown that one efficient means of communicating the signal across a channel is as shown in Figure 1.2.[2] The signal is first converted from analog to digital (A/D), then modulated with a high frequency carrier and transmitted from an antenna. Likewise, at the receiver, the modulated signal is received by an antenna, demodulated, and then converted back from digital to analog format (D/A). In DSP, we are primarily focused on baseband signals, i.e., the A/D and D/A conversion stages.
- *Analyzing the analog signal for use in a voice recognition telephone system.* Today, it is common to encounter voice recognition systems when we call businesses, railway and airline reservation lines, banks, and a host of other places. In these systems, the analog signal is sampled and then analyzed, using a Fast Fourier Transform (FFT), as shown in Figure 1.3. The analyzed frequency spectrum data is input into a microprocessor, which matches the latter spectrum with a stored signal spectrum. If a good match is determined between the two signal spectra, then a particular operation is initiated.

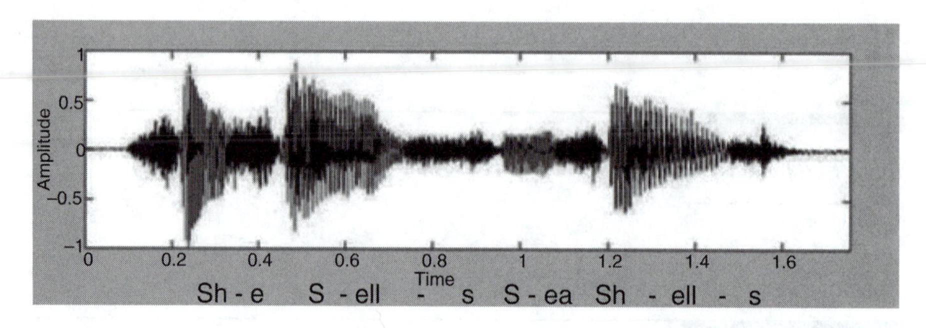


**FIGURE 1.1**
Time waveform of "She sells sea shells."

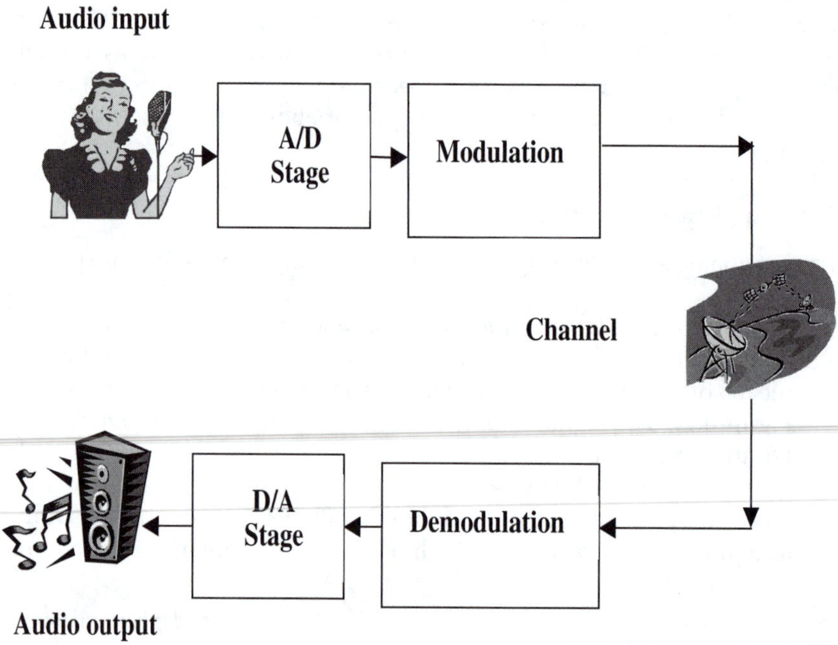

Audio input

A/D Stage

Modulation

Channel

D/A Stage

Demodulation

Audio output

**FIGURE 1.2**
Block diagram of digital communications system.

- *Analyzing the analog signal to obtain useful information.* This application is useful for biomedical signals, such as EKG (heart) and EEG (brain). In such a case, the analog signal is sampled and then analyzed using a FFT. The FFT spectrum of the EKG, for example, can reveal several useful parameters about the patient, such as high potassium levels (hyperkalemia) or low potassium levels (hypokalemia).

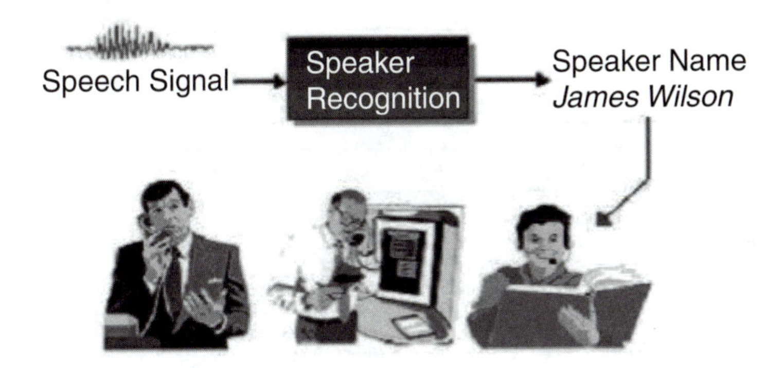


**FIGURE 1.3**
Voice recognition system.

### 1.1.2 Discrete-Time Signals and Systems

One common factor in the applications listed above is the A/D conversion, which will be discussed in detail in Chapter 4. However, the fundamental operation in the A/D process is the process of sampling, as shown in Figure 1.4. The continuous time signal $x(t)$ is sampled uniformly at $T$ = 0.1 second to yield a *discrete-time signal* $x(nT)$ or just $x(n)$, $N_1 \le n \le N_2$. Some common examples of discrete-time signals are shown in Figure 1.5

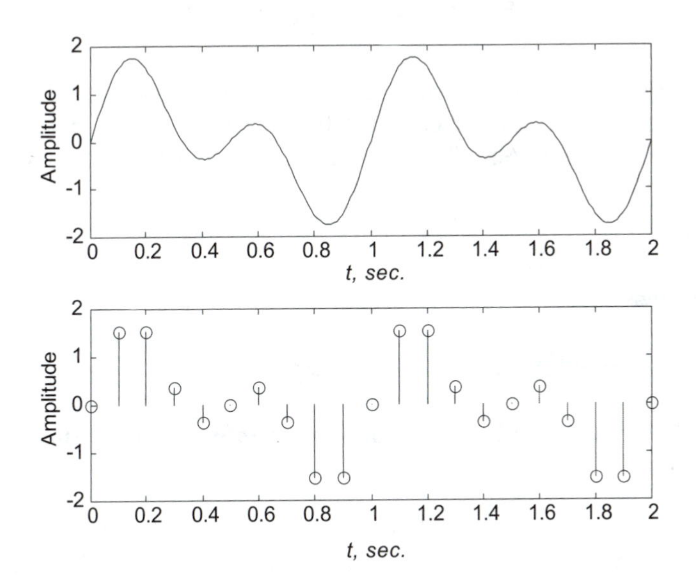


**FIGURE 1.4**
Sampling process with $T$ = 0.1 s.

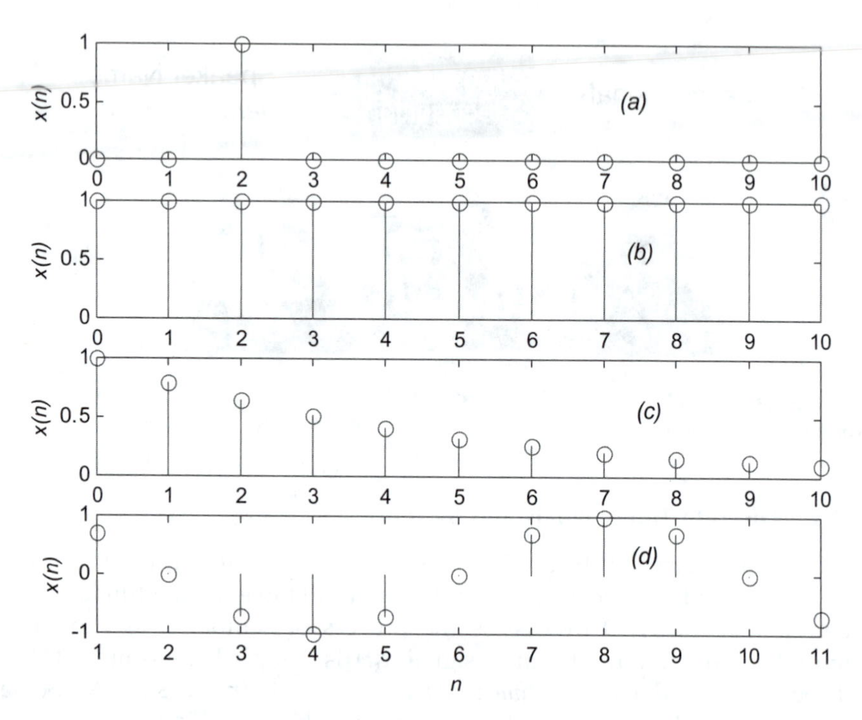


**FIGURE 1.5**
Examples of common discrete-time signals. (a) Unit impulse sequence x(n) = δ(n-2), (b) Unit step sequence x(n) = u(n), (c) Exponential signal $x(n) = 0.8^n u(n)$, (d) Sinusoidal sequence x(n) = cos(πn/4).

Likewise, systems that operate on discrete-time signals are termed *discrete-time systems.* One important class of these systems is called a *linear time invariant* (*LTI*) system. We will discuss discrete-time LTI systems in detail in Chapter 2.

## 1.2 Problem Solving

**Exercise 1: Solve the following problems, briefly outlining the important steps:**

a. DSP primarily deals with commonly occurring signals such as speech, music, video, EKG (heart), and EEE (brain). Research the Internet to get information on the approximate frequency ranges of the five signals listed above.

b. Approximately sketch the following discrete-time signals in the range $0 \le n \le 10$:
   i. $x(n) = u(n) - u(n-3)$
   ii. $x(n) = u(3-n)$
   iii. $x(n) = 0.5^n [u(n) - u(n-5)]$

c. Determine if each of the following signals is periodic or nonperiodic. If a signal is periodic, specify its fundamental period:
   i. $x(n) = e^{j\pi n}$
   ii. $x(n) = 0.7^n u(n)$
   iii. $x(n) = \cos(\pi n/2)\cos(\pi n/4)$

## 1.3 Computer Laboratory: Introduction to MATLAB®/Simulink®

Programming software utilized in Digital Signal Processing (DSP) applications can be placed in the following two categories:

- **Simulation software:** This software is used to model DSP systems and, hence, is a very valuable tool to design actual systems. In this laboratory, MATLAB and Simulink® are used to model systems. While MATLAB requires programs to be written, Simulink® is a graphical tool, which has built-in DSP blocks.
- **Software for hardware control:** This software is required to run DSP hardware such as Digital Signal Processors (DSPs). Examples of this software are DSP routines written in Assembly or C programming languages.

### 1.3.1 MATLAB Basics

Please try out each of the commands given below and familiarize yourself with the types of MATLAB commands and formats.[3]

**System Operating Commands**

PC-based MATLAB can be opened by clicking on the MATLAB icon. The MATLAB prompt is >>, which indicates that commands can be started either line by line or by running a stored program. A complete program, consisting of a set of commands, can be stored in a MATLAB file for repeated use as follows:

- Open a file in any text editor (either in MATLAB or otherwise) and write the program.
- After writing the program, exit the program after saving it as a **filename.m** file.
- To execute the program, either run the file from the MATLAB text editor or type the filename after the prompt:

  >> filename

The program will run, and the results and error messages, if any, will be displayed on the screen. Plots will appear on a new screen.

*Numbers*

**Example:** Generate the real numbers $z_1 = 3$, $z_2 = 4$.

```
>> z1 = 3
>> z2 = 4
```

**Example:** Generate the complex numbers $z_1 = 3+j4$, $z_2 = 4+j\,5$.

```
>> z1 = 3+j*4
>> z2 = 4+j*5
```

*Note:* The symbol $i$ can be used instead of $j$ to represent $\sqrt{-1}$.

**Example:** Find the magnitude and phase of the complex number $z = 3 + j4$.

```
>> z = 3+j*4
>> zm = abs(z); gives the magnitude of z
>> zp = angle(z); gives the phase of z in radians
```

Addition or subtraction of numbers (real or complex):

```
>> z = z1+z2; addition
>> z = z1-z2; subtraction
```

Multiplication or division of numbers (real or complex):

```
>> z = z1*z2; multiplication
>> z = z1/z2; division
```

*Vectors and Matrices*

**Example:** Generate the vectors $x$ = [1 3 5] and $y$ = [2 0 4 5 6].

```
>> x = [1 3 5]; generates the vector of length 3
>> y = [2 0 4 5 6]; generates the vector of length 5
```

Addition or subtraction of vectors $x$ and $y$ of same length:

```
>> z = x+y; addition
>> z = x-y; subtraction
```

Multiplication or division of vectors $x$ and $y$ of same length:

```
>> z = x. * y; multiplication
>> z = x./y; division
```

*Note:* The dot after x is necessary since x is a vector and not a number.

### *Creating one-dimensional and two-dimensional vector spaces using MATLAB*

The command:

>> `x = linspace(x1,x2,N);` generates N points between $x_1$ and $x_2$; and stores it in the vector $x$

The commands:

>> `x = linspace(x1,x2,N1);` generates $N_1$ points between $x_1$ and $x_2$; and stores it in the vector $x$

>> `y = linspace(y1,y2,N2);` generates $N_2$ points between $y_1$ and $y_2$; and stores it in the vector $y$

>> `[X,Y] = meshgrid(x,y);` generates the two-dimensional meshgrid [X,Y]

### *Programming with vectors*

Programs involving vectors can be written using either FOR loop or vector commands. Since MATLAB is basically a vector-based program, it is often more efficient to write programs using vector commands. However, FOR loop commands give a clearer understanding of the program, especially for the beginner:

**Example:** Sum the following series:

S = 1 + 3 + 5 … … + .99.

- FOR loop approach:

```
>> S = 0.0; initializes the sum to zero
>> for i = 1 : 2 : 99
S = S + i
end
>> S; gives the value of the sum
```

- Vector approach:

```
>> i =1: 2 : 99; creates the vector i
>> S = sum (i); obtains the sum S
```

**Example:** Generate the discrete-time signal $y(n) = n\ sin(\pi n/2)$ in the interval $0 \leq n \leq 10$.

- FOR loop approach:

```
>> for n = 1:1: 11
n1(n) = n - 1
y(n) = n1(n) * sin(pi*n1(n)/2)
end
>> y; gives the vector y
>> stem(n1,y); plots the signal y(n) vs. n with impulses
```

- Vector approach:

```
>> n = 0 : 10; creates the vector n
>> y = n.*sin(pi*n/2); obtains the vector y
>> stem(n,y); plots the signal y(n) vs. n with impulses
```

***Basic Signal Operations In MATLAB***

**Example:** Define the discrete-time signal $x(n) = n\,u(n)$ in a vector in the range $0 \leq n \leq 10$ and plot the signal.

***Solution:***

```
>> n = 0:10; defines the vector n of length 11
>> u (1:11) = ones(1,11); defines the unit step vector u of length 11
>> x = n. *u; defines the product n u(n)
>> stem(n,x); plots the discrete signal x(n)
```

**Exercise 2: Working with vectors and matrices**

- Write a MATLAB program to sketch the following discrete-time signals in the time range of $-10 \leq n \leq 10$. Please label all the graph axes clearly. If the sequence is complex, plot the magnitude and angle separately.
  i. $x(n) = u(n) - u(n-3)$
  ii. $x(n) = \sin(\pi n/3)\,u(n)$
  iii. $x(n) = 0.5^n\,e^{j\pi n/2}$
- There are two main forms of vector or matrix multiplication. In MATLAB, if two vectors, $a$ and $b$ (both vectors of size $1xN$) are given, then the two possible MATLAB multiplication commands are: $y$ = a*b and $y$ = a.*b.

i. Comment on the differences between these two commands, and clearly state what would be the outputs of these two operations.
ii. Write a small MATLAB program to evaluate these two commands for the case of $a$ = [1 2 3] and $b$ = [4 5 6]. Are both the operations a*b and a.*b possible? If not, what change in the syntax would make the operation possible?

### 1.3.2 Simulink® Basics

After logging into MATLAB, you will receive the prompt >>. In order to open Simulink®,[3] type in the following:

```
>> simulink
```

Alternately, click on the Simulink® icon in the MATLAB Command window.

**General Simulink® Operations**

Two windows will open: the **model window** and the **library window**. The model window is the space used for creating your simulation model. In order to create the model of the system, components will have to be selected from the Simulink® library, using the computer mouse, and dragged into the model window. If you browse the library window, the following sections will be seen. Each section can be accessed by clicking on it.

- **Sources:** This section consists of different signal sources, such as sinusoidal, triangular, pulse, random, or files containing audio or video signals.
- **Sinks:** This section consists of measuring instruments such as scopes and displays.
- **Math:** This section consists of linear components performing operations such as summing, integration, and product.
- **Continuous:** This section consists of simulation blocks to simulate continuous-time systems.
- **Discrete:** This section consists of simulation blocks to simulate discrete-time systems.
- **Signal Routing:** Multiplexers, demultiplexers.
- **Blocksets:** Blocksets specify different specialized areas of electrical engineering. Some examples are given below:
    - Communications
    - DSP
    - Neural Nets
    - CDMA Spread Spectrum models

**Editing, Running, and Saving Simulink® Files**

The complete system is created in the model window by using components from the various available libraries. Once a complete model is created, save the model into a **filename.mdl** model file. Click on **Simulation** and select Run. The simulation will run, and the output plots can be displayed by clicking on the appropriate **sinks**. Save the output plots, also, into files. The model and output files can be printed out from the files.

**Demo Files**

Try out the **demo** files, both in the main library window and in the **Toolboxes** window. There are several illustrative demonstration files in the areas of signal processing, image processing, and communications.

**Exercise 3: Simulation of continuous-time LTI systems**

a. Create Simulink® models for the continuous-time system shown in Figure 1.6. Before starting any simulation, select the **Simulation** button from the model window, and then select the **Parameter** button. Modify the **start time** and **stop time** of the simulation to complete at least two periods of the signal source.
b. Run the simulation for the sinusoidal signal, $x(t)$, with amplitude of 5 Volts and frequency of $\omega = 10$ rad./s. The signal $n(t)$ is a pseudo-random noise with maximum amplitude of 0.5 volts. Obtain a printout of the combined output signal $y(t)$ on the time scope, and familiarize yourself with the settings.

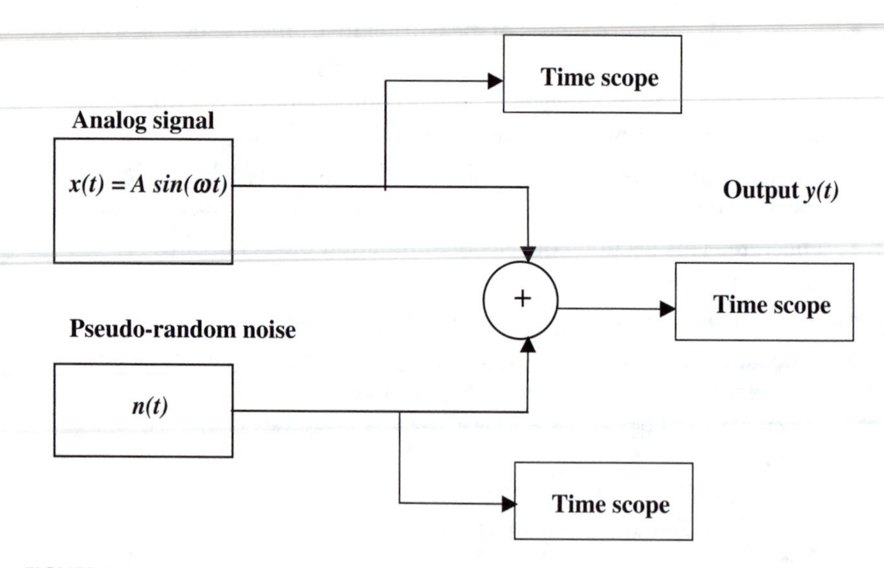


**FIGURE 1.6**
Simulink® model of a continuous-time system.

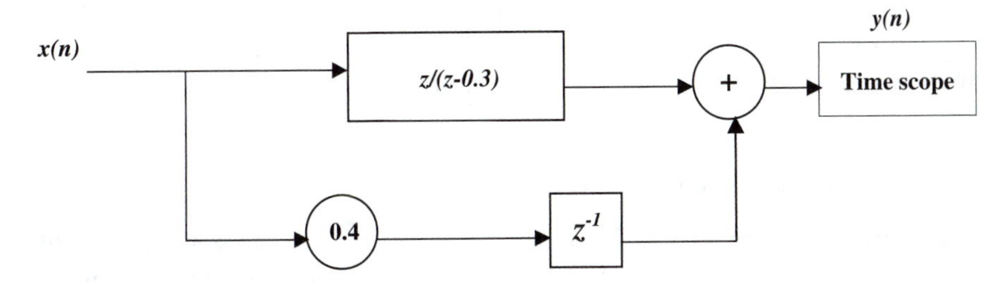


**FIGURE 1.7**
Simulink® model of a discrete-time system.

c. Change sinusoidal signal amplitude (2V, 10V) and frequency (20 rad./s, 50 rad./s), and obtain the printouts of the output on the time scope.

**Exercise 4: Simulation of discrete-time LTI systems**

a. Create Simulink® models for the discrete-time system shown in Figure 1.7. Please note that in all discrete-time simulation blocks, the appropriate sampling time, $T$, sec. should be specified if required.
b. Obtain a printout of the output signal on the time scope, for an input signal of $x(t) = 3\cos(2\pi t/5)$, in the time range $0 \le t \le 5$ sec., when the signal is sampled at a time interval of $T = 0.5$ sec.
c. Change the input signal amplitude to 6 volts and the input signal frequency to twice its original value; observe the output on the time scope. Obtain a printout of the output signal. Comment on the differences between output signals obtained in part (b) and part (c) of this exercise.

## 1.4 Hardware Laboratory: Working with Oscilloscopes, Spectrum Analyzers, Signal Sources

Hardware equipment used in DSP applications can be classified into three main categories.

### 1.4.1 Sinks or Measuring Devices

Sinks or measuring devices are used to accurately graph input signals in two domains: *time* and *frequency*. The HP 54510A 100 MHz Digitizing Oscilloscope measures the amplitude and frequency of signals as a function of time, whereas the HP 8590L RF Spectrum Analyzer measures the spectrum of the input signal as a function of frequency. Please see Appendix D and Appendix B, respectively, for manufacturer's specifications and operating instructions on these two pieces of measuring equipment.

### 1.4.2 Dynamic Signal Analyzers

These analyzers are more advanced equipment, which can generate regular signals, as well as random noise. They can measure signals in both time and frequency and can also measure frequency responses of devices. The HP 35665 Dynamic Signal Analyzer in the laboratory is a multipurpose piece of equipment. See Appendix C for manufacturer's details.

### 1.4.3 Sources

Sources generate signals that vary in shape, amplitude, frequency, and phase. The source used in the laboratory is the Agilent HP 3324A Synthesized Sweep Generator. See Appendix A for the manufacturer's specifications and other operating instructions.

## 1.5 Digital Signal Processors (DSPs)

Digital signal processors are very widely used components in many communications systems, such as cell phones. They are, essentially, programmable microprocessors that can perform many signal processing functions such as filtering, mathematical operations, convolution, and Fast Fourier Transform (FFT). The Texas Instruments TMS320C6711 floating point DSP Starter Kit will be used widely in the laboratory to study important functions such as filtering and noise removal. See Appendix E for manufacturer's details.

**Exercise 5: Basic experiments using sources and measuring equipment**

- In this experiment, basic time and frequency measurements will be performed using the oscilloscope and signal analyzer.
  a. Connect the equipment together as shown in the schematic in Figure 1.8. Use BNC cables and a BNC Tee to connect the circuit. Make sure the HP3324A Synthesized sweep generator power button is in the **off** position.
  b. Set the sweep generator to output a sinusoidal signal, with an amplitude of 5 volts and frequency $f = 2$ MHz. Observe the time-domain signal output on the oscilloscope and note down the measured amplitude and frequency of the sinusoidal signal.

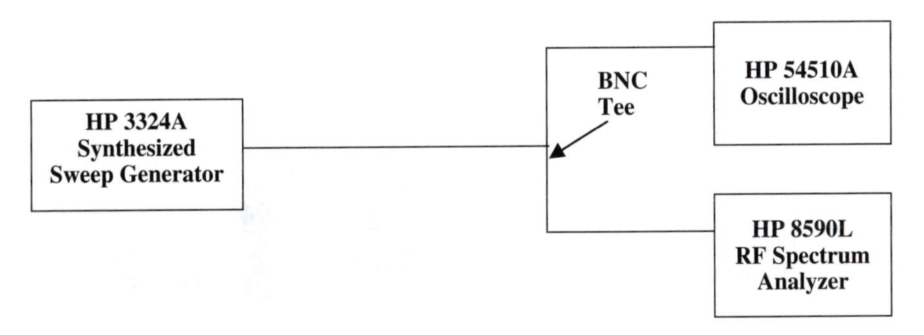


**FIGURE 1.8**
Experimental setup for signal time and frequency analysis.

c. Set the signal analyzer to a start frequency of 1.0 MHz and a stop frequency of 3.0 MHz. Observe the frequency-domain output on the Spectrum Analyzer and note down the measured amplitude and frequency of the sinusoidal signal. Use the markers in the Signal Analyzer to peak search mode to track the peak value in the signal spectrum.

d. Comment on the differences, if any, between the set sweep generator frequency, oscilloscope output frequency, and signal analyzer output frequency.

- In this experiment, time and frequency measurements will be performed using the HP 35665 Dynamic Signal Analyzer. The Dynamic Signal Analyzer is a very versatile low-frequency equipment that can analyze and manipulate signals in the frequency range of 0 to 50 kHz.

  a. The Dynamic Signal Analyzer has one output port called **source**, and two input ports called **channel 1** and **channel 2**. The output of the source port is controlled by the source key on the top section of the Signal Analyzer. The source can generate several kinds of sources including single frequency sinusoidal, swept frequency sinusoidal, and random noise.

  b. Connect the source output to channel 1 input with a BNC cable. Select the source key, and select sinusoidal source with a frequency of 10 kHz and an amplitude of 5 volts. Select the **measurement** key, and alternate between time and frequency settings to observe the signal in both the domains. Time and frequency plots of the signal can be viewed simultaneously by using the **dual channel** display mode.

  c. Repeat part (b) of this experiment with a **random signal** source having a peak amplitude of 1 volt. Observe the random signal in both time and frequency domains.

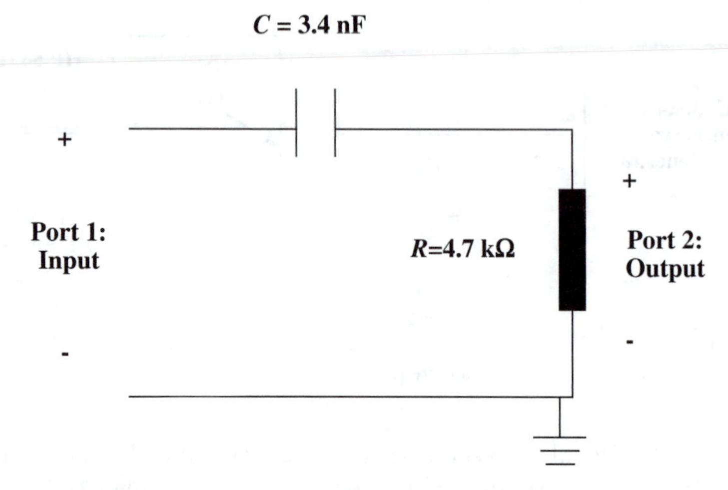


**FIGURE 1.9**
Low-pass filter circuit with 10 kHz cutoff frequency.

- In this experiment, frequency response measurements will be performed using the HP 35665 Dynamic Signal Analyzer.
  a. The Dynamic Signal Analyzer can measure the frequency response of a passive device, such as an electrical filter, for example, in the frequency range of 0 to 50 kHz, In this experiment, we will determine the frequency response of a low-pass filter, having a cutoff frequency of 10 kHz.
  b. Build the filter circuit shown in Figure 1.9, with port 1 as the *input port* and port 2 as the *output port*. Since the circuit is quite simple, it can be put together even on a breadboard for testing. However, a permanent circuit soldered together would be more ideal.
  c. Connect the filter circuit to the HP 35665 Signal Analyzer as shown in Figure 1.10. The source output of the Signal Analyzer is simultaneously connected to the input port of the filter circuit and also to **channel 1** (Ch1) of the Signal Analyzer, while the output port of the filter circuit is connected to **channel 2** (Ch2) of the Signal Analyzer.
  d. It is important to set the Signal Analyzer settings appropriately to obtain the frequency response of the filter circuit. Select the **frequency** key on the Signal Analyzer, and set the **start** and **stop** frequencies to 0 Hz and 20 KHz, respectively. Select the **measure** key, and set the measurement to **2 channel** measurements, and then select **frequency response** setting. Select the **source** key, and set the source to **chirp** signal, which will generate a swept

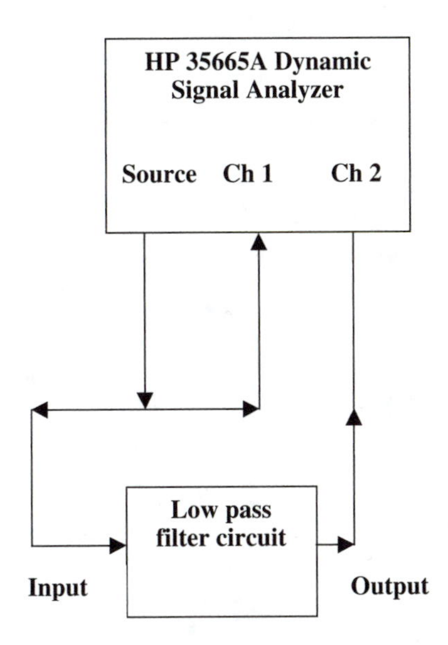


**FIGURE 1.10**
Experimental setup for frequency response measurement.

frequency signal for 0 to 20 kHz. Set the amplitude of the chirp signal to 1 volt.

e. Select the **source** key, and set it to the **on** position. Finally select the **scale** key, and set it at **autoscale**. The frequency response of the filter should now appear on the screen of the signal analyzer.

## References

1. Brigham, E., *Fast Fourier Transform and Its Applications,* Pearson Education, Toronto, 1988.
2. Lathi, B.P., *Modern Digital and Analog Communication Systems,* 3rd Edition, Oxford University Press, New York, 1998.
3. *Student Edition of MATLAB/Simulink,* Mathworks, Natick, MA.
4. Oppenheim, A.V and Schafer, R.W., with Buck, J.R., *Discrete-Time Signal Processing,* 2nd Edition, Prentice Hall, Upper Saddle River, NJ, 1998.
5. Kumar, B.P., *Digital Signal Processing Laboratory,* California State University, Sacramento, 2003.

# 2

# *Discrete-Time Signals and Systems*

## 2.1 Brief Theory of Discrete-Time Signals and Systems

In the previous chapter, we defined the concepts of discrete-time signals and systems and noted that an important class of these systems is the LTI (linear time invariant) system. It is well known that linearity implies superposition, and time-invariance implies that properties of the system do not change with time. Two of the most important concepts associated with discrete-time LTI systems are *linear convolution* and *linear constant coefficient* difference equations.

**Linear Convolution**

*Linear convolution* is a natural process of LTI systems. It defines the input-output relation of the system and is defined as:

$$y(n) = x(n) * h(n) \tag{2.1}$$

where the * symbol denotes the convolution process, $x(n)$ is the system input, $y(n)$ is the sytem output, and $h(n)$ is the *impulse response* of the system. The impulse response is the output of the system, when the input $x(n) = \delta(n)$, the unit impulse. The actual convolution process is defined as:

$$y(n) = \sum_{k=-\infty}^{\infty} x(k)h(n-k) \tag{2.2}$$

for all values of $n$. Convolution can be performed using the *sliding tape method,* as shown below or, more practically, by using computer software, as will be described in Section 2.3.

**Example**

Determine the linear convolution of two discrete-time sequences, $x(n)$ and $h(n)$, given by:

$$x(n) = [1\ 1\ 1\ 1]$$

$$h(n) = [1\ 2\ 3]$$

***Solution***

The linear convolution is given by:

$$y(n) = \sum_{k=-\infty}^{\infty} x(k)h(n-k)$$

*Sliding tape method:* This method can be done by hand calculation, if the number of points in both the sequences is quite small.[1] The procedure is as follows:

- Write the sequences $x(m)$, $h(m)$, and $h(-m)$ as shown below. The sequence $h(-m)$ is obtained by mirroring the sequence $h(m)$ about the $m = 0$ axis. Then the dot product of the vectors $x(m)$ and $h(-m)$ gives the convolution output $y(0)$. Similarly, the next term in the table below, $h(1\text{-}m)$, is obtained by shifting $h(-m)$ by *one* step to the right. The dot product of the vectors $x(m)$ and $h(1\text{-}m)$ gives the convolution output $y(1)$. The process is continued until the output $y(n)$ remains at zero.

$$
\begin{aligned}
x(m) &= [0\ \ 0\ \ 0\ \ 1\ \ 1\ \ 1\ \ 1] \\
h(m) &= [0\ \ 0\ \ 0\ \ 1\ \ 2\ \ 3\ \ 0] \\
h(0\text{-}m) &= [0\ \ 3\ \ 2\ \ 1\ \ 0\ \ 0\ \ 0];\ y(0) = 1 \\
h(1\text{-}m) &= [0\ \ 0\ \ 3\ \ 2\ \ 1\ \ 0\ \ 0];\ y(1) = 3 \\
h(2\text{-}m) &= [0\ \ 0\ \ 0\ \ 3\ \ 2\ \ 1\ \ 0];\ y(2) = 6 \\
h(3\text{-}m) &= [0\ \ 0\ \ 0\ \ 0\ \ 3\ \ 2\ \ 1];\ y(3) = 6 \\
h(4\text{-}m) &= [0\ \ 0\ \ 0\ \ 0\ \ 0\ \ 3\ \ 2];\ y(4) = 5 \\
h(5\text{-}m) &= [0\ \ 0\ \ 0\ \ 0\ \ 0\ \ 0\ \ 3];\ y(5) = 3
\end{aligned}
$$

Any more shift in the sequence $h(m)$ will result in a zero output. Hence, the output vector is:

$$y(n) = [1\ 3\ 6\ 6\ 5\ 3].$$

Note that the length of the output vector y(n) = [length of $x(n)$]+ [length of $h(n)$] –1:

$$\text{Length of } y(n) = 4 + 3 - 1 = 6.$$

This is a general law of discrete-time linear convolution.

### *Linear Constant Coefficient Difference Equations*

The difference equation is another fundamental relation between the input and output of LTI systems. It describes the discrete-time process executed by the system, such as a high-pass or a low-pass filtering. The difference equation takes the following form:

$$a_0 y(n) + a_1 y(n-1) + \ldots a_N y(n-N) = b_0 x(n) + b_1 x(n-1) + \ldots b_M x(n-M)$$

or compactly as follows:

$$\sum_{k=0}^{N} a_k y(n-k) = \sum_{k=0}^{M} b_k x(n-k) \tag{2.3}$$

Important properties of the difference equations are:

- The order of the difference equation is $N$, where $N \geq M$.
- The coefficients $a_i$, $i = 0, 1, 2, \ldots N$, and $b_i$, $i = 0, 1, 2, \ldots M$ are constant, real numbers and define the properties of the system. For example, one set of coefficients could generate a low-pass filter, while a different set of coefficients could generate a highpass filter. The latter aspect shows the great simplicity of digital systems.

### 2.1.1 Introduction to Z-Transforms and the System Function *H*(*z*)

In the previous section, we introduced two fundamental properties of LTI systems, linear convolution and difference equation. The link between these two fundamental aspects of LTI systems is provided by the Z-transform. The Z-transform is a very important time-to-frequency domain transformation of the basic discrete-time signal $x(n)$. The Z-transform is defined by the equation:

$$X(z) = \sum_{n=-\infty}^{\infty} x(n) z^{-n} \tag{2.4}$$

where the variable $z$ is complex and the function $X(z)$ is defined in the complex plane.

A concise list of Z-transform properties is given in Table 2.1. Utilizing property (2) from Table 2.1, in Equation 2.2, we get the transformed equation:

$$Y(z) = X(z)\, H(z) \tag{2.5}$$

**TABLE 2.1**

Z-Transform Theorems

| Property | f(n) | F(z) |
|---|---|---|
| Linearity | $A\,x_1(n) + B\,x_2(n)$, for constants $A$, $B$ | $A\,X_1(z) + B\,X_2(z)$ |
| Convolution | $x(n) * h(n)$ | $X(z)\,H(z)$ |
| Time Shift | $x(n - n_0)$ | $X(z)\,z^{-n}_{n.}$ |
| Frequency Scaling | $z_0{}^n x(n)$ | $X(z/z_0)$ |
| Differentiation | $n\,x(n)$ | $-z\dfrac{dX(z)}{dz}$ |

Similarly, using property (3) from Table 2.1 in Equation 2.3, we get the transformed equation:

$$Y(z)\sum_{k=0}^{N} a_k z^{-k} = X(z)\sum_{k=0}^{M} b_k z^{-k} \tag{2.6}$$

Comparing Equation 2.5 and Equation 2.6, we get the following *system function* $H(z)$:

$$\frac{Y(z)}{X(z)} = H(z) = \frac{\sum_{k=0}^{M} b_k z^{-k}}{\sum_{k=0}^{N} a_k z^{-k}} \tag{2.7}$$

**Properties of the System Function *H(z)***

The system function or transfer function $H(z)$ can be expressed in the following forms

- *Polynomial form*

This form can be obtained by expanding Equation 2.7 to yield:

$$H(z) = \frac{b_0 + b_1 z^{-1} + b_2 z^{-2} + \ldots\, b_M z^{-M}}{a_0 + a_1 z^{-1} + a_2 z^{-2} + \ldots\, a_N z^{-N}} \tag{2.8}$$

- *Pole-zero form and stability of the system*

This form can be obtained by factorizing the numerator and denominator polynomials in Equation 2.8 to yield:

$$H(z) = \frac{b_0\left(1 - c_1 z^{-1}\right)\left(1 - c_2 z^{-1}\right)\ldots\left(1 - c_M z^{-1}\right)}{a_0\left(1 - d_1 z^{-1}\right)\left(1 - d_2 z^{-1}\right)\ldots\left(1 - d_N z^{-1}\right)} \tag{2.9}$$

The $M$ roots of the numerator polynomial $c_1\ c_2 \ldots c_M$ are the *zeros*, and the $N$ roots of the denominator polynomial $d_1\ d_2 \ldots d_N$ are the *poles* of the system function.

The poles of the system function define the *stability* of the system in the complex $z$-plane. If all the poles of the system function lie inside the unit circle, hence satisfying $|d_k| < 1$, for $k = 1, 2, \ldots N$, then the system is *unconditionally stable*. For an example system function

$$H(z) = \frac{(z-2)}{z(z-0.5)(z-0.8)}$$

the poles, or the roots of the denominator polynomial, are located at $z = 0$, 0.5 and 0.8. Since all the poles lie inside the unit circle, the system $H(z)$ is unconditionally stable.

### 2.1.2 System Frequency Response $H(e^{j\omega})$

The frequency response of the system is very important to define the practical property of the system, such as low-pass or high-pass filtering. It can be obtained by considering the system function $H(z)$ on the unit circle shown in Figure 2.1, which corresponds to the $z = e^{j\omega}$ circle. Hence, from Equation 2.7, the frequency response is:

$$H(e^{j\omega}) = H(z)\big|_{z=e^{j\omega}} \tag{2.10}$$

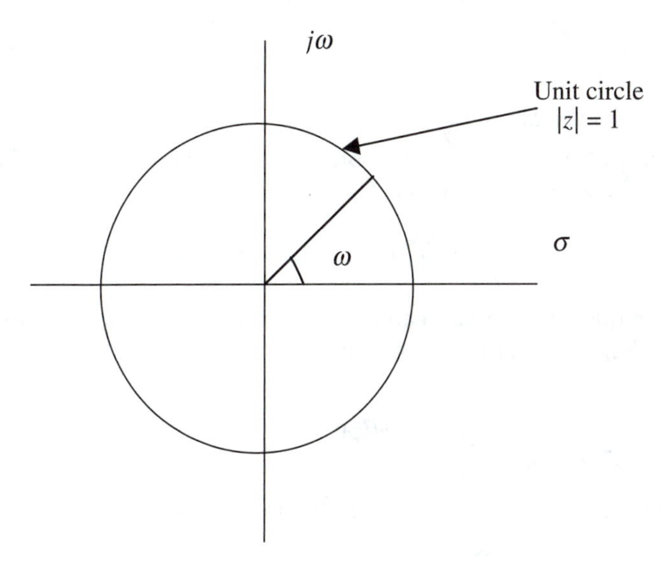


**FIGURE 2.1**
Complex $z$-plane with the unit circle.

Since the function $e^{j\omega}$ is periodic with period $2\pi$ radians, the frequency response, $H(e^{j\omega})$, of any discrete-time system is also periodic with period $2\pi$ radians. This is one important distinction between continuous-time and discrete-time systems. The frequency response is, in general, complex, and hence we define the *magnitude response* $|H(e^{j\omega})|$ and phase response $/H(e^{j\omega})$.

**Example**

An $N$th order *comb filter* has the system function $H(z) = 1 - z^{-N}$.

i. Determine the pole-zero locations of $H(z)$.
ii. Evaluate the impulse response $h(n)$.
iii. Determine the frequency response (magnitude and phase) of the system.

***Solution***

The system function is

$$H(z) = 1 - z^{-N}$$

$$= \frac{z^N - 1}{z^N}$$

i. From the system function, the poles and zeros can be obtained:

*Zeros*: $N$ zeros at

$$z^N = 1 = e^{j2\pi k}, \quad k = 0, 1, \ldots N-1$$

$$\Rightarrow z = e^{j2\pi k/N}, \quad k = 0, 1, \ldots N-1$$

*Poles:* $N$ poles at $z = 0$.

ii. The impulse response can be obtained by taking the inverse Z-transform of the system function $H(z)$:

$$h(n) = (n)\ h(n) = \delta(n) - \delta(n-1)$$

iii. The frequency response can also be obtained from the system function $H(z)$ by using Equation 2.10:

$$H\left(e^{j\omega}\right) = H(z)\Big|_{z=e^{j\omega}} = 1 - e^{-j\omega N}$$

which can be simplified to yield the final result:

$$H\left(e^{j\omega}\right) = 2je^{-j\omega N/2} \sin\left(\omega N/2\right)$$

The magnitude and phase responses can be obtained from the frequency response as follows:

*Magnitude response:* $\left|H\left(e^{j\omega}\right)\right| = 2\left|\sin\left(\omega N/2\right)\right|$

*Phase response:* $\left[H\left(e^{j\omega}\right)\right] = 90° + \left[\sin\left(\omega N/2\right)\right] - \left(\omega N/2\right)\frac{180}{\pi}$, degrees

Since the frequency response of the system $H(e^{j\omega})$ is periodic in $2\pi$ radians, it is sufficient to define both the magnitude and phase reponses of the system in the interval $-\pi \leq \omega \leq \pi$ radians.

### 2.1.3 Important Types of LTI Systems

The fundamental properties of LTI systems directly affect the behavior of practical electrical components such as filters, amplifiers, oscillators, and antennas. Some commonly used systems are described briefly below.

- ***Inverse system:*** As the name implies, the inverse system $H_i(z)$ of a given system $H(z)$ is defined as:

$$H_i(z) = \frac{1}{H(z)} \tag{2.11}$$

  Inverse systems are used in audio and video processing to recover signals coming through noisy channels. However, *the inverse system may not be stable, even if the original system is stable.* This is because the zeros of the system $H(z)$ are the poles of the system $H_i(z)$. In order to overcome this problem, we would require a *minimum-phase system.*

- ***All-pass system:*** As the name implies, an all-pass system has a frequency response magnitude that is independent of $\omega$. A stable system function of the form:

$$H_{ap}(z) = \frac{z^{-1} - a^*}{1 - az^{-1}} \tag{2.12}$$

  has the frequency response:

$$\begin{aligned} H_{ap}\left(e^{j\omega}\right) &= \frac{e^{-j\omega} - a^*}{1 - ae^{-j\omega}} \\ &= e^{-j\omega}\frac{1 - a^* e^{j\omega}}{1 - ae^{-j\omega}} \end{aligned} \tag{2.13}$$

  which implies that the magnitude response $\left|H_{ap}(e^{j\omega})\right| = 1$.

- ***Minimum-phase system:*** A minimum-phase system has both its poles and zeros inside the unit circle. This implies that *both a minimum-phase system and its inverse are stable.* Hence, in audio and video processing units, the inverse system can be designed as the reciprocal of a minimum-phase system as follows:

$$H_i(z) = \frac{1}{H_{\min}(z)} \tag{2.14}$$

This will ensure that the inverse system is stable. For any specific rational system function $H(z)$, the minimum-phase system $H_{min}(z)$ exists and can be derived using the theorem that $H(z) = H_{\min}(z)\, H_{ap}(z)$, as shown in the example below.

**Example:** Specify the minimum-phase system $H_{\min}(z)$ for the following system function $H(z)$:

$$H(z) = \frac{\left(1+2z^{-1}\right)\left(1-\frac{1}{2}z^{-1}\right)}{z^{-1}\left(1+\frac{1}{3}z^{-1}\right)}$$

such that

$$\left|H\left(e^{j\omega}\right)\right| = \left|H_{\min}\left(e^{j\omega}\right)\right|.$$

***Solution:***

The solution is given in a series of steps below.

- Rewrite the system function $H(z)$ using $z$ terms instead of $z^{-1}$ terms.

$$H(z) = \frac{(z+2)\left(z-\frac{1}{2}\right)}{\left(z+\frac{1}{3}\right)}$$

- Identify the zeros and poles of $H(z)$ that lie outside the unit circle i.e., $|z| > 1$. (*These zeros and poles are highlighted in* **bold**.)

  Poles: $z = -1/3$

  Zeros: $\mathbf{z = -2}$, $z = 1/2$

- Rewrite $H(z)$ as follows, by replacing every root outside the unit circle with its conjugate reciprocal root (i.e., replace $z = c$ with $z = 1/c^*$, where the symbol * denotes complex conjugate):

$$H(z) = \frac{\left(z+\frac{1}{2}\right)\left(z-\frac{1}{2}\right)}{\left(z+\frac{1}{3}\right)} \quad \frac{(z+2)}{\left(z+\frac{1}{2}\right)}$$

which can be separated as $H(z) = H_{\min}(z) \times H_{ap}(z)$.

- In order to ensure $|H(e^{j\omega})| = |H_{\min}(e^{j\omega})|$, we have to determine the magnitude of the all-pass system response as follows:

$$H_{ap}(z) = \frac{(z+2)}{\left(z+\frac{1}{2}\right)}$$

- Hence, the frequency response of the all-pass system is

$$H_{ap}\left(e^{j\omega}\right) = \frac{\left(e^{j\omega}+2\right)}{\left(e^{j\omega}+\frac{1}{2}\right)}$$

and the magnitude response is $|H(e^{j\omega})| = |H(e^{j0})|$ (*since the system is all-pass, the magnitude response is independent of frequency*)

$$\left|H_{ap}\left(e^{j\omega}\right)\right| = \frac{\left(e^{j0}+2\right)}{\left(e^{j0}+\frac{1}{2}\right)} = \frac{(1+2)}{\left(1+\frac{1}{2}\right)} = 2$$

- Hence, if we redefine

$$H_{\min}(z) = 2\frac{\left(z+\frac{1}{2}\right)\left(z-\frac{1}{2}\right)}{\left(z+\frac{1}{3}\right)}$$

and

$$H_{ap}(z) = \frac{1}{2}\frac{(z+2)}{\left(z+\frac{1}{2}\right)}$$

then the condition

$$\left|H\left(e^{j\omega}\right)\right| = \left|H_{\min}\left(e^{j\omega}\right)\right|$$

is ensured.

## 2.2 Problem Solving

**Exercise 1: Solve the following problems, briefly outlining the important steps.**

a. Determine the overall impulse response of the system shown below in Figure 2.2, where

$$h_1(n) = \delta(n-1) + 3\delta(n)$$

$$h_2(n) = \delta(n-2) + 2\delta(n)$$

$$h_3(n) = 6\delta(n-6) + 7\delta(n-4) - 3\delta(n-1) + \delta(n)$$

b. A causal, linear, time-invariant discrete-time system has system function:

$$H(z) = \frac{\left(1-0.5z^{-1}\right)\left(1+4z^{-2}\right)}{\left(1-0.64z^{-2}\right)}$$

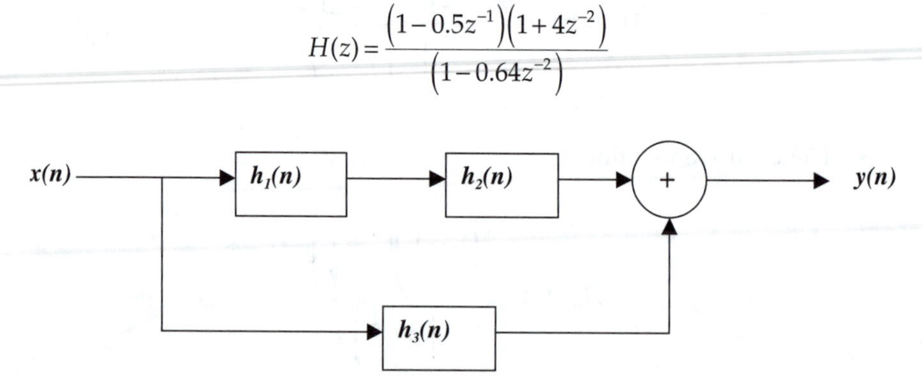


**FIGURE 2.2**
Figure for problem (a).

i. Find expressions for a minimum-phase system $H_1(z)$ and an all-pass system $H_{ap}(z)$ such that:

$$H(z) = H_1(z) H_{ap}(z)$$

ii. Plot the pole-zero plots of $H(z)$, $H_1(z)$, and $H_{ap}(z)$.

c. A simple model for multipath channel is described by the difference equation:

$$x(n) = s(n) - e^{-8\alpha} s(n-8)$$

We wish to recover $s(n)$ from $x(n)$ with a linear time-invariant system. Find the causal and stable system function $H(z) = Y(z)/X(z)$ such that its output $y(n) = s(n)$.

d. Consider a causal LTI system described by the difference equation:

$$y(n) = p_0\, x(n) + p_1\, x(n-1) - d_1\, y(n-1)$$

where $x(n)$ and $y(n)$ denote, respectively, its input and output.

Determine the difference equation of its *inverse* system.

## 2.3 Computer Laboratory: Simulation of Continuous Time and Discrete-Time Signals and Systems

This section consists of examples in MATLAB,[3] followed by the laboratory exercises. Please test the example problems, before proceeding to the exercises.

### MATLAB Examples

**Example:** Solve the following difference equation for $0 \le n \le 10$:

$$y(n) = y(n-1) + 2\, y(n-2) + x(n-2)$$

given that $x(n) = 4 \cos(\pi\, n/8)$, $y(0) = 1$ and $y(1) = 1$.

***Solution:***

```
>> y=[1 1]
>> x(1)=4
```

```
>> x(2)=4*cos(pi/8)
>> for n = 3:11
           n1= n-1
           x(n)  = 4*cos(pi*n1/8)
           y(n)  = y(n-1) + 2*y(n-2) + x(n-2)
   end
>> y; displays the vector y(n)
>> stem(y); plots the vector y(n)
>> xlabel('n'); defines the x-axis on plot
>> ylabel('y(n)'); defines the y-axis on plot
>> title('system output y(n)'); defines the title of
the plot
```

**Example:** Find the system output $y(n)$, $0 \le n \le 10$, of a LTI system when the input $x(n) = (0.8)^n [u(n) - u(n-5)]$ and the impulse response $h(n) = (0.5)^n [u(n) - u(n - 10)]$. Plot the vectors $\boldsymbol{x}$, $\boldsymbol{h}$, and $\boldsymbol{y}$ on the same page using *subplot* commands.

***Solution:***

- FOR loop approach

```
>> for n = 1:10
            n1(n)= n-1
            h(n)  = (0.5)^n1(n)
   end
>> for n = 1:5
            n2(n)= n-1
            x(n)  = (0.8)^n2(n)
   end
>> y=conv(x, h); performs the convolution of the
vectors x and h
>> n4 = size(n1)+size(n2)-1; calculates the size of
the output vector y
>> n3 = 0:n4-1
>> subplot(3,1,1); divides the page into 3 rows and 1
column format
>> stem(n1,x); plots the input vector x
>> subplot(3,1,2)
>> stem(n2,h); plots the impulse response vector h
```

```
>> subplot(3,1,3)
>> stem(n3,y); plots the output vector y
```

- **Vector approach**

  The vector approach is a more compact and more efficient from of MATLAB programming

```
>> n1 = 0:4
>> x = 0.8. ^ n1
>> n2 = 0:9
>> h = 0.5. ^ n2
>> y = conv(x,h)
>> k1 = size(n1) + size(n2)-1
>> k = 0:k1-1
>> subplot(3,1,1); divides the page into 3 rows and
1 column format
>> stem(n1,x); plots the input vector x
>> subplot(3,1,2)
>> stem(n2,h); plots the impulse response vector h
>> subplot(3,1,3)
>> stem(k,y); plots the output vector y
```

NOTE: **The output vector $y$ will be of length 14. In general, if the vector $x$ is of length $N$, and the vector $h$ is of length $M$, then the output vector $y$ is of length $N + M - 1$.**

**Exercise 2: Plotting of continuous-time and discrete-time signals**

a. Plot the following continuous-time signals in the range $-5 \leq t \leq$ 5 seconds.
   i. a.$x(t) = 5 \sin(10\ t) + 10 \sin(20\ t)$
   ii. b.$x(t) = 2\ e^{-(a\ t^2)}$, $a = 0.1$

b. Plot the following discrete-time signals in the range $-5 \leq n \leq 5$.
   i. a.$x(n) = 0.8^n\ u(n)$
   ii. b.$x(n) = [\sin(0.1\pi n)]/\pi n$

**Exercise 3: Discrete-time convolution**

Find the system output $y(n)$, $0 \leq n \leq 10$, of a LTI system when the input $x(n) = \delta(n) + 3\ \delta(n-1) + 4\ \delta(n-3)$, and the impulse response $h(n) = (0.5)n\ [u(n) - u(n-5)]$. Write a concise MATLAB program, using vector approach to model the output, $y(n)$, of the system, for the given input. Plot the vectors $\boldsymbol{x}$, $\boldsymbol{h}$, and $\boldsymbol{y}$ on the same page using subplot commands.

**Exercise 4: Creation of system models in MATLAB**

The MATLAB model of an LTI system can be created using the **zpk** command. For example, consider the LTI system:

$$H(z) = \frac{\left(1-0.5z^{-1}\right)\left(1+4z^{-2}\right)}{\left(1-0.64z^{-2}\right)}$$

a. Converting the system function in terms of $z$, we obtain:

$$H(z) = \frac{(z-0.5)\left(z^2+4\right)}{z\left(z^2-0.64\right)}$$

b. The MATLAB command for system model creation is as follows:

```
>> z = zpk('z',Ts); H = (z-0.5)*(z^2+4)/(z*(z^2+0.64))
```

where *Ts* is the sampling frequency of the discrete-time system in seconds. Then the model *H* represents the system function *H*(*z*). If we require the pole-zero plot, for example, we could give the command:

```
>> pzmap(H)
```

which would generate the pole-zero plot of the system. In a similar fashion the following command:

```
>> [p,z]=pzmap(H)
```

returns vectors ***p*** and ***z***, which contain the poles and zeros, respectively, of the system *H*.

In addition, the frequency response of the system can also be obtained from the model *H* by using the following two commands:

```
>> w = 0:dw:pi; sets the frequency axis ω = [0 π] in steps of dω.

>> bode(H,w); plots the Bode magnitude and phase plots of the system H
```

Alternately, the command

```
>>[mag phase] = bode(H,w);generates magnitude and phase vectors of the system H
```

Write a compact MATLAB program to create a system model H, corresponding to the LTI system given below:

$$H(z) = \frac{\left(1+2z^{-1}\right)\left(1-\frac{1}{2}z^{-1}\right)}{z^{-1}\left(1+\frac{1}{3}z^{-1}\right)}$$

c. Extend the program to complete the following:

- Generate the pole-zero plot of the system $H(z)$.
- Determine the poles and zeros of the system function $H(z)$, and have the program automatically generate the poles and zeros of the corresponding minimum-phase system $H_{min}(z)$.
- Create the system model $H_{min}$ corresponding to the minimum-phase system $H_{min}(z)$, and generate the pole-zero plot of the minimum-phase system.
- Create the system model $H_{ap}$ corresponding to the all-pass system $H_{ap}(z)$, and generate the pole-zero plot of the all-pass system.
- Generate the Bode plots (magnitude and phase) of the minimum-phase system $H_{min}(z)$ and the all pass system $H_{ap}(z)$.

**Exercise 5: Simulation of discrete-time LTI systems using MATLAB and Simulink[2]**

Simulate the following discrete-time systems, shown in Figure 2.3(a), Figure 2.3(b), and Figure 2.3(c) in both MATLAB (using time-domain) and Simulink (using frequency or Z-domain). Please provide the program listing/block diagrams and required output plots.

a. In each case, obtain the output of the overall system, $y(n)$, with an input $x(n) = u(n) - u(n - 4)$.
b. Plot the output, $y(n)$, in the range $-10 \leq n \leq 10$. Please label the plots clearly.

*Hint:* In order to represent LTI systems in Simulink, use the discrete filter block from the Simulink library. In this block, enter the coefficients of the numerator and denominator polynomials of the LTI system function $H(z)$.

**Exercise 6: Image files and two-dimensional data matrices[3]**

A number of image files can be accessed from the MATLAB directory or from the Internet. Download the image as a .jpg file into your local PC directory using the following command:

```
>> y = imread ('filename.jpg'); loads the image into
the matrix y
```

and the commands:

```
>> imagesc(y); plots the image in a new window
>> colorbar; attaches a color scale to the figure
```

a. Access any two-dimensional image file from the directory. Example: *coneplot.jpg, light_ex2.jpg, surface_ex2.jpg.*

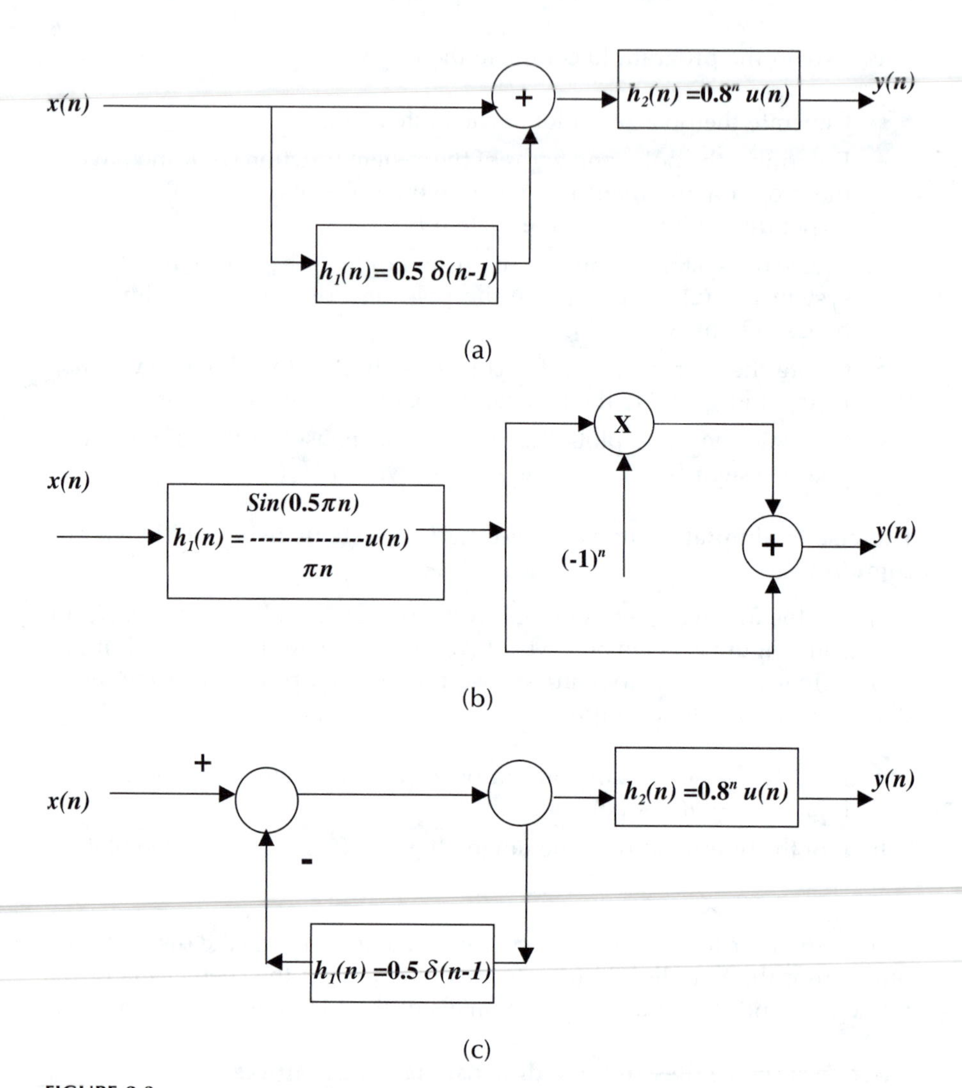


**FIGURE 2.3**
(a) Discrete–time system; (b) discrete–time system; (c) discrete–time system.

i. In each case, print the corresponding image on the network printer.
ii. Each image is stored as a matrix $a(m,n)$. Determine the size of the matrix for each image and determine the number of pixels $(m \times n)$ in each case.
iii. Determine the maximum and minimum values of the image matrices for each image.

b. Generate *surf* plots of the following two-dimensional functions $f(x,y)$, in the range of $-2 \le x \le 2$ ; $-2 \le y \le 2$, using 100 points along both $x$ and $y$ axes:

i. $f(x, y) = x^2 + y^2$
ii. $f(x, y) = e^{-(|x|^2 - |y|^2)}$

In each case, appropriately list all axes labels.

## References


1. Lathi, B.P., *Linear Signals and Systems*, Oxford University Press, New York, 2001.
2. Haddad, R.A. and Parsons, T.W., *Digital Signal Processing — Theory, Applications, and Hardware*, Computer Science Press, 1991.
3. *Student Edition of MATLAB/Simulink*, Mathworks, Natick, MA, Version 5.3, 1999.
4. Oppenheim A.V and Schafer, R.W., with Buck, J.R., *Discrete-Time Signal Processing*, 2nd Edition, Prentice Hall, Upper Saddle River, NJ, 1998.
5. Kumar, B.P., *Digital Signal Processing Laboratory*, California State University, Sacramento, 2003.

[illegible]

to [illegible] [illegible] [illegible] all [illegible]

## References

1. Lathi, B.P., Linear Systems and Signals, [illegible] University Press, New York, [illegible]
2. [illegible] and [illegible], J.W., Digital Signal [illegible], [illegible] Science Press, 199[illegible].
3. [illegible]
4. [illegible] Third Edition, [illegible]
5. [illegible] Signal Processing [illegible] 20[illegible].

# 3

# Time and Frequency Analysis of Discrete-Time Signals

## 3.1 Brief Theory of Discrete-Time Fourier Transform, Discrete Fourier Transform, and Fast Fourier Transform

In the previous chapter, the Z-transform was shown to be an effective tool in linking the time and frequency domains of a discrete-time signal $x(n)$. However, in order to specify practical properties of discrete-time systems, such as low-pass filtering or high-pass filtering, it is necessary to transform the complex $z$-plane to the real-frequency, $\omega$, axis. Specifically, the region of the complex $z$-plane that is used in this transformation is the unit circle, specified by the region $z = e^{j\omega}$. The resulting transform is the *Discrete-Time FourierTransform* (*DTFT*), which will be discussed first in this chapter.

Due to the need for a more applicable and easily computable transform, the *Discrete Fourier Transform* (*DFT*) was introduced, which is very homogeneous in both forward (time to frequency) and inverse (frequency to time) formulations. The crowning moment in the evolution of DSP came when the *Fast Fourier Transform* (*FFT*) was discovered by Cooley and Tukey in 1965.[1,2] The FFT, which is essentially a very fast algorithm to compute the DFT, makes it possible to achieve real-time audio and video processing.

### 3.1.1 Discrete-Time Fourier Transform

The frequency response of the system is very important in defining the practical property of the system, such as low-pass or high-pass filtering. It can be obtained by considering the system function $H(z)$ on the unit circle, as discussed in Chapter 2. Similarly, for a discrete-time sequence $x(n)$, we can define the Z-transform $X(z)$ on the unit circle as follows:

$$X(e^{j\omega}) = X(z)\Big|_{z=e^{j\omega}} = \sum_{n=-\infty}^{\infty} x(n)e^{-j\omega n} \tag{3.1}$$

**TABLE 3.1**

DTFT Theorems

| Property | f(n) | F(ω) |
|---|---|---|
| Periodicity | $x(n)$ | $X(\omega) = X(\omega + 2m\pi)$, for integer $m$ |
| Convolution | $x(n) * h(n)$ | $X(\omega)\, H(\omega)$ |
| Time Shift | $x(n - n_0)$ | $X(\omega)\, e^{-j\omega\, n0}$ |
| Frequency Shift | $e^{j\omega\, n0}x(n)$ | $X(\omega - \omega_0)$ |
| Time Reversal | $x(-n)$ | $X(-\omega)$ |

The function $X(e^{j\omega})$ or $X(\omega)$ is also called the *Discrete-Time Fourier Transform* (*DTFT*) of the discrete-time signal $x(n)$. The inverse DTFT is defined by the following integral:

$$x(n) = \frac{1}{2}\pi \int_{-\pi}^{\pi} X(\omega)e^{j\omega n}d\omega \tag{3.2}$$

for all values of $n$. The significance of the integration operation in Equation 3.2 will be clear after discussing the *periodicity* property of the DTFT in the next section.

### *Properties of Discrete-Time Fourier Transform*

A concise list of DTFT properties is given in Table 3.1.

*Analog frequency and digital frequency*

The fundamental relation between the analog frequency, $\Omega$, and the digital frequency, $\omega$, is given by the following relation:

$$\omega = \Omega T \tag{3.3a}$$

or alternately,

$$\omega = \Omega/f_s \tag{3.3b}$$

where $T$ is the sampling period, in sec., and $f_s = 1/T$ is the sampling frequency in Hz.

This important transformation will be discussed more thoroughly in Chapter 5. Note, however, the following interesting points:

- The unit of $\Omega$ is radian/sec., whereas the unit of $\omega$ is just radians.
- The analog frequency, $\Omega$, represents the *actual physical frequency of the basic analog signal*, for example, an audio signal (0 to 4 kHz) or a video signal (0 to 4 MHz). The digital frequency, $\omega$, is the transformed frequency from Equation 3.3a or Equation 3.3b and can be considered as a mathematical frequency, corresponding to the digital signal.

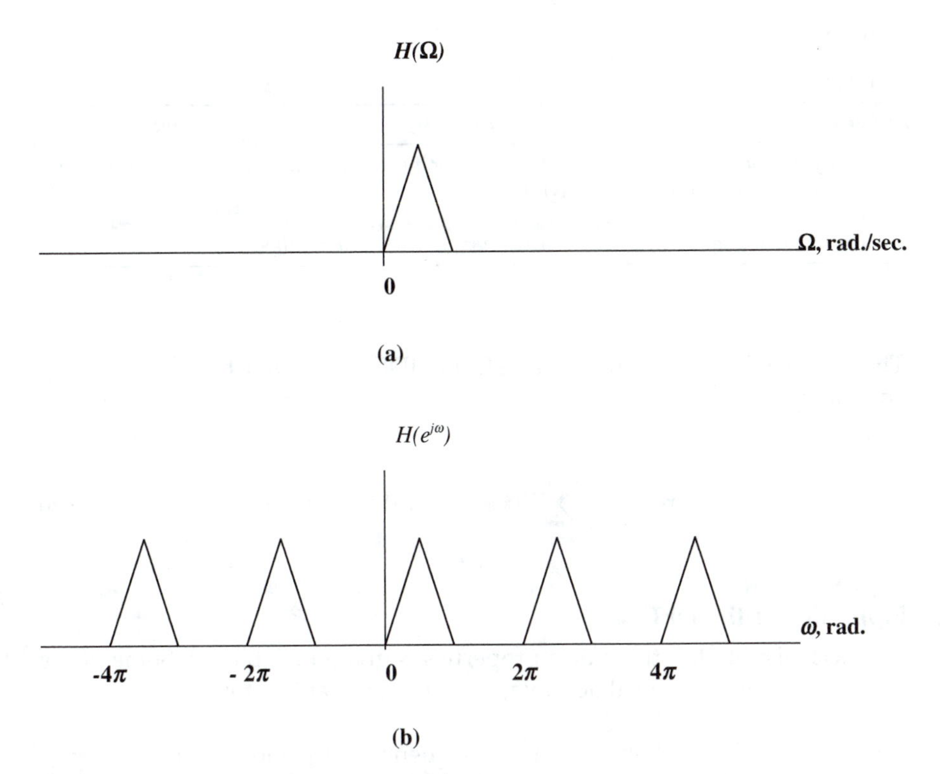


**FIGURE 3.1**
(a) Analog frequency response and (b) digital frequency response.

### *Analog frequency response and digital frequency response*

One of the most important differences between discrete-time systems and analog systems is that discrete-time systems have a periodic frequency response, $H(e^{j\omega})$, while analog systems have a nonperiodic Fourier transform $H(j\Omega)$. Figure 3.1 illustrates this difference in between $H(j\Omega)$ and $H(e^{j\omega})$.

### 3.1.2 Discrete Fourier Transform

The Discrete Fourier Transform (DFT) is a practical extension of the DTFT, which is discrete in both time and the frequency domains. As discussed in the previous section, the DTFT $X(\omega)$ is a periodic function with period $2\pi$ radians. This property is used to the divide the frequency interval $(0, 2\pi)$ into $N$ points, to yield the DFT of the discrete-time sequence $x(n)$, $0 \le n \le N - 1$ as follows:

$$X(k) = X(\omega)\Big|_{\omega=2\pi k/N} = \sum_{n=0}^{N-1} x(n)e^{-j2\pi nk/N}, \; 0 \le k \le N-1 \tag{3.4}$$

**TABLE 3.2**
DFT Theorems

| Property | f(n) | F(k) |
|---|---|---|
| Periodicity in $n$ and $k$ | $x(n) = x(n \pm mN)$, for integer $m$ | $X(k)=X(k \pm mN)$, for integer $m$ |
| $N$-point circular convolution | $x(n)$ Ⓝ $h(n)$ | $X(k)\ H(k)$ |
| Circular time shift | $x((n - n_0)_N)$ | $X(k)\ e^{-j2\pi n_0 k/N}$ |
| Circular frequency shift | $e^{j2\pi nk_0/N}\ x(n)$ | $X((k - k_0)_N)$ |

The Inverse Discrete Fourier Transform (IDFT) is given by the following equation:

$$x(n) = \frac{1}{N}\sum_{k=0}^{N-1} X(k)e^{j2\pi nk/N},\ 0 \le n \le N-1 \tag{3.5}$$

## Properties of the DFT

A concise list of DFT transform properties is given in Table 3.2. Some of the key features and practical advantages of the DFT are as follows:

- The DFT maintains the time sequence $x(n)$ and the frequency sequence $X(k)$ as finite vectors having the same length $N$. Additionally, as seen from Equation 3.4 and Equation 3.5, the DFT and IDFT are both finite sums, which makes it very convenient to program these equations on computers and microprocessors.
- *The time-frequency relation* is a very important relation in practical DFT applications. The index $n$ corresponds to the time value $t = n\Delta t$, sec., where $\Delta t$ is the sampling time interval. The index $k$ corresponds to the frequency value $\omega = k\Delta\omega$, radians, where $\Delta\omega$ is the DFT output frequency interval. Then, for a given $N$-point DFT, the time frequency relation is given by

$$\Delta\omega = 2\pi/(N\ \Delta t)$$

- The concept of *time shift* in the DFT is defined circularly: the sequence $x(n)$, $0 \le n \le N - 1$ is represented at $N$ equally spaced points around a circle as shown in Figure 3.2a, for $N = 8$. Then, a circular shift, represented as $x((n - 5)_8)$, for example, is implemented by moving the entire sequence $x(n)$ counter-clockwise by five points, as illustrated in Figure 3.2b. Hence, the sequence $x(n) = [x(0)\ x(1)\ x(2)\ x(3)\ x(4)\ x(5)\ x(6)\ x(7)]$, and the shifted sequence $x((n - 5)_8) = [x(3)\ x(4)\ x(5)\ x(6)\ x(7)\ x(0)\ x(1)\ x(2)]$.

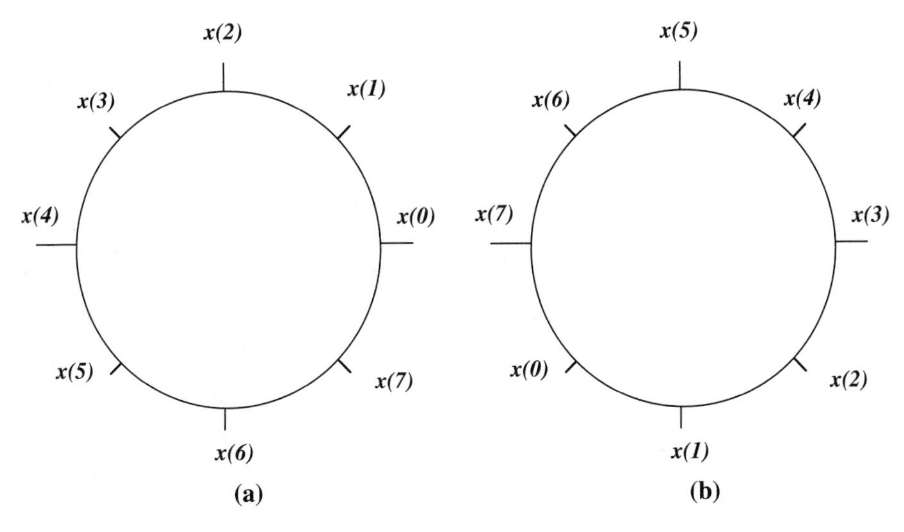


**FIGURE 3.2**
(a) Sequence $x(n)$ and (b) circularly shifted sequence $x((n-5)_8)$.

### Circular Convolution

An $N$-point circular convolution of two sequences $x(n)$ and $h(n)$ is defined as:

$$h(n) = x_1(n) \circledN h(n) = \sum_{m=0}^{N-1} x(m)h\big((n-m)_N\big), \; 0 \le n \le N-1 \tag{3.6}$$

*Note:* The sequences $x(n)$, $h(n)$, and $y(n)$ have the same vector length of $N$.

### Example

Determine the circular convolution of the two 8-point discrete-time sequences, $x_1(n)$ and $x_2(n)$, given by

$$x_1(n) = x_2(n) = \begin{cases} 1, & 0 \le n \le 4 \\ 0, & 5 \le n \le 7 \end{cases}$$

Discuss the different methods of performing circular convolution.

### *Solution*

The 8-point circular convolution is given by

$$x(n) = x_1(n) \circledast_8 x_2(n) = \sum_{m=0}^{7} x_1(m)x_2\big((n-m)_8\big)$$

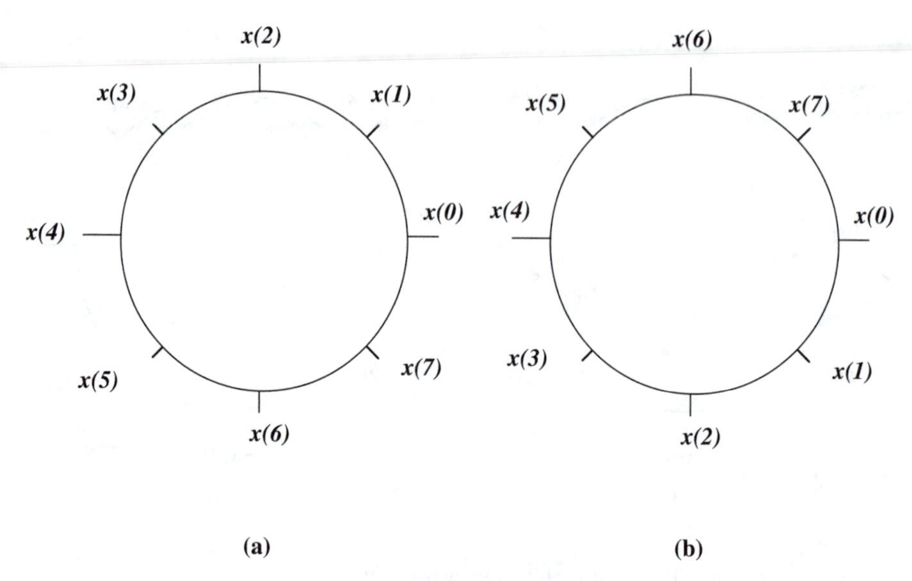


**FIGURE 3.3**
(a) Sequence $x(m)$ and (b) reflected sequence $x((-m)_8)$.

Circular convolution can be carried out either by *analytic techniques*, such as the sliding tape method, or by *computer techniques*, such as MATLAB. We will discuss both approaches below.

The *sliding tape method* can be done by hand calculation, if the number of points in the DFT, $N$, *is quite small.* The procedure is as follows:

- Write the sequences $x_1(m)$, $x_2(m)$, and $x_2((-m)_8)$ as shown below.

| | | |
|---|---|---|
| $x_1(m)$ | = [1 1 1 1 1 0 0 0] | |
| $x_2(m)$ | = [1 1 1 1 1 0 0 0] | |
| $x_2((-m)_8)$ | = [1 0 0 0 1 1 1 1]; | $x(0) = 2$ |
| $x_2((1-m)_8)$ | = [1 1 0 0 0 1 1 1]; | $x(1) = 2$ |
| $x_2((2-m)_8)$ | = [1 1 1 0 0 0 1 1]; | $x(2) = 3$ |
| $x_2((3-m)_8)$ | = [1 1 1 1 0 0 0 1]; | $x(3) = 4$ |
| $x_2((4-m)_8)$ | = [1 1 1 1 1 0 0 0]; | $x(4) = 5$ |
| $x_2((5-m)_8)$ | = [0 1 1 1 1 1 0 0]; | $x(5) = 4$ |
| $x_2((6-m)_8)$ | = [0 0 1 1 1 1 1 0]; | $x(6) = 3$ |
| $x_2((7-m)_8)$ | = [0 0 0 1 1 1 1 1]; | $x(7) = 2$ |

The sequence $x_2(-m)$ is obtained from the sequence $x_2(m)$ by writing the first element in the vector $x_2(m)$, then starting with the last element in $x_2(m)$ and continuing *backwards*. Then, the dot product of the vectors $x_1(m)$ and $x_2((-m)_8)$ gives the convolution output $x(0)$. Similarly, the next term in the sequence, $x_2((1 - m)_8)$, is obtained by shifting $x_1(-m)$ by one step to the right, and *back again to the beginning of the vector.* The dot product of the vectors $x_1(m)$ and $x_2((1 - m)_8)$ gives the convolution output $x(1)$.

- Alternately, one could arrange the vector elements $x_1(m)$ and $x_2(m)$ in $N = 8$ equally spaced points around a circle, as shown in Figure 3.3a. The vector $x_2((-m)_8)$ is obtained by reflecting the vector elements of $x_2(m)$ about the horizontal axis as shown in Figure 3.3b. The vector $x_2((1 - m)_8)$ is obtained by shifting the elements of the vector $x_2(m)$ by one position counter-clockwise around the circle. Hence, the output vector is $x(n) = [2\ 2\ 3\ 4\ 5\ 4\ 3\ 2]$.

Using the *computer method*, the circular convolution of the two sequences, $x_1(n)$ and $x_2(n)$, can also be obtained by using the convolution property of the DFT, which is listed as Property 2 in Table 3.2 above. This method consists of three steps.

- **Step 1:** Obtain the 8-point DFTs of the sequences $x_1(n)$ and $x_2(n)$:

$$x_1(n) \to X_1(k)$$

$$x_2(n) \to X_2(k)$$

- **Step 2:** Multiply the two sequences $X_1(k)$ and $X_2(k)$:

$$X(k) \to X_1(k)X_2(k), \text{ for } k = 0, 1, 2 \ldots 7.$$

- **Step 3:** Obtain the 8-point IDFT of the sequence $X(k)$, to yield the final output $x(n)$:

$$X(k) \to x(n), \text{ for } n = 0, 1, 2 \ldots 7.$$

A brief MATLAB program to implement the procedure above is given below:

```
% MATLAB Program for Circular Convolution
clear;
x1=[1 1 1 1 1 0 0 0]  ; sequence x1(n)
x2=[1 1 1 1 1 0 0 0]  ; sequence x2(n)
X1=fft(x1)            ; DFT of x1(n)
X2=fft(x2)            ; DFT of x2(n)
X=X1.*X2              ; DFT of x(n)
x=ifft(X)             ; IDFT of X(k)
```

*Note:* MATLAB automatically utilizes a radix-2 FFT if N is a power of 2. If N is not a power of 2, then it reverts to a non-radix-2 process. The FFT process will be explained in the next section.

### 3.1.3 The Fast Fourier Transform

The Fast Fourier Transform, or the FFT, as it is popularly termed, is probably the single most famous computer program in the field of electrical engineering and represents the most practical version of the DTFT, which is what we initially started out with. It is, essentially, a much faster computation method of the DFT, discussed in the previous section. The exceptional computational efficiency of the FFT is achieved by using some periodic properties of the exponential functions in Equation 3.4 and Equation 3.5. An example FFT computation of a 4-point DFT will be outlined below.

**FFT Computation of a 4-point DFT**

We can define a 4-point DFT of the sequence $x(n) = [x(0)\ x(1)\ x(2)\ x(3)]$ from Equation 3.4 as follows:

$$X(k) = \sum_{n=0}^{3} x(n) e^{-j2\pi nk/4}, \; 0 \le k \le 3 \tag{3.7}$$

or using compact notation:

$$X(k) = \sum_{n=0}^{3} x(n) W_4^{nk}, \quad 0 \le k \le 3 \tag{3.8}$$

where, in general

$$W_N^{nk} = e^{-j2\pi nk/N}$$

Expanding Equation 3.8, we have:

$$X(k) = x(0)W_4^{0k} + x(1)W_4^{1k} + x(2)W_4^{2k} + x(3)W_4^{3k}, \quad 0 \le k \le 3 \tag{3.9}$$

In order to compute $X(k)$ for each value of $k$, from Equation 3.9, we would require, in general, four complex multiplications. Hence, for all four values of $k$, we would require a total of 4 x 4 = 16 complex multiplications, *if the DFT were computed directly from Equation* 3.9. Now, we will show in a series of steps, how the FFT reduces the latter multiplication count.

**Step 1: Dividing the computation into even and odd index terms**

Grouping the even and odd index terms from Equation 3.9, we have:

$$X(k) = x(0)W_4^{0k} + x(2)W_4^{2k} + x(1)W_4^{1k} + x(3)W_4^{3k}, 0 \le k \le 3 \tag{3.10}$$

Factorizing the odd terms we have

$$\begin{aligned} X(k) &= x(0)W_4^{0k} + x(2)W_4^{2k} + W_4^{1k}\left\{x(1) + x(3)W_4^{2k}\right\},\ 0 \le k \le 3 \\ &= \quad X_1(k) \quad + \quad W_4^{1k}\ X_2(k),\ 0 \le k \le 3 \end{aligned} \tag{3.11}$$

**Step 2: Computation of $X_1(k)$**

$$X_1(k) = x(0)W_4^{0k} + x(2)W_4^{2k},\ 0 \le k \le 3 \tag{3.12}$$

Examination of Equation 3.12 yields the following interesting relations:

$$X_1(2) = X_1(0) = x(0) + x(2)$$

$$X_1(3) = X_1(1) = x(0) - x(2)$$

Hence, we have to calculate, only the following terms:

$$X_1(0) = x(0)W_4^{00} + x(2)W_4^{20} \tag{3.13a}$$

and

$$X_1(1) = x(0)W_4^{01} + x(2)W_4^{21} \tag{3.13b}$$

**Step 3: Computation of $X_2(k)$**

$$X_2(k) = x(1) + x(3)W_4^{2k},\ 0 \le k \le 3 \tag{3.14}$$

Examination of Equation 3.14 yields similar interesting relations:

$$X_2(2) = X_2(0) = x(1) + x(3)$$

$$X_2(3) = X_2(1) = x(1) - x(3)$$

Hence, we have to calculate, only the following terms:

$$X_2(0) = x(1) + x(3)W_4^{20} \tag{3.15a}$$

and

$$X_2(1) = x(1) + x(3)W_4^{21} \tag{3.15b}$$

**Step 4: Computation of the product $W_4^{1k} X_2(k)$**

This is the final step of the FFT computation. Again, there are some surprising relations:

$$W_4^{10} X_2(0) = -W_4^{12} X_2(2) = X_2(0)$$

$$W_4^{11} X_2(1) = -W_4^{13} X_2(3) = j\, X_2(1)$$

Hence, we have to calculate, only the following terms:

$$W_4^{10} X_2(0) \tag{3.16a}$$

and

$$W_4^{11} X_2(1) \tag{3.16b}$$

Now we can make a multiplication count from Steps 2, 3, and 4. Step 2 requires four complex multiplications, by using Equation 3.13(a) and Equation 3.13(b). Step 3 requires two complex multiplications from Equation 3.15(a) and Equation 3.15(b), and finally Step 4 requires two complex multiplications, as seen from Equation 3.16(a) and Equation 3.16(b). Hence, we get a total multiplication count of eight using the FFT procedure, as compared with the direct computation of the DFT from Equation 3.9, which requires 16 multiplications. The latter reduction in multiplication count will be generalized into a formula in the next section.

### *Properties of the FFT*

Some key properties of the FFT are given below:

- An $N$-point DFT or $N$-point IDFT requires $N^2$ complex multiplications if computed directly from Equation 3.4 and Equation 3.5. However, the same computation can be done with only $N\, Log_2 N$ complex multiplications, when a *radix*-2 ($N$ is a power of 2) FFT is used. This is especially significant for large values of $N$: when $N = 128$, the number of complex multiplications is 16384 for direct computation of DFT and only 896 for a *radix*-2 FFT computation.
- FFT algorithms also exist when $N$ is not a power of 2. These algorithms are called *non-radix*-2 FFT.

### *Practical Usage of the FFT: Computation of Fast Fourier Transform with MATLAB*

The FFT was a major breakthrough in the efficient and fast computation of the Fourier transform of speech, music, and other fundamental signals. However, while the FFT is a very general formulation, there are some important points to keep in mind, when utilizing the FFT on *periodic* and *nonperiodic* signals.

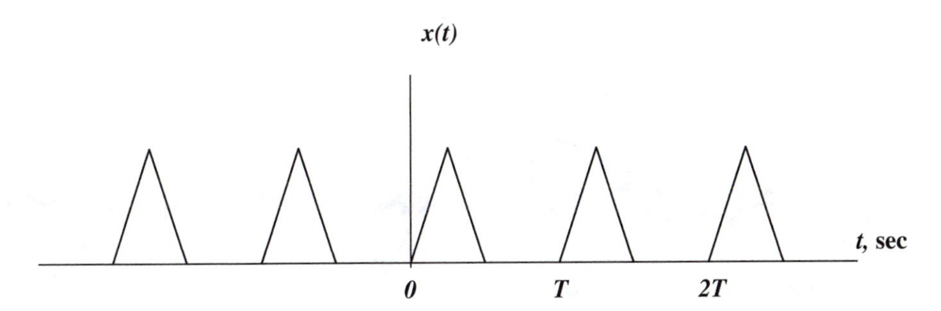


**FIGURE 3.4**
Periodic signal $x(t)$.

*FFT evaluation of periodic signals*

**Step 1.** Sample the signal $x(t)$, shown in Figure 3.4 *over* 1 *period* of the signal, $T = 2\pi/\omega_0$, where $\omega_0$ is the angular frequency of the signal. The sampling interval is:

$$\Delta t = T/N$$

where $N$ is the number of points in the FFT.

**Step 2.** Generate the sampled signal $x(n)$, $n = 0, 1, \dots N-1$. The input signal is stored as a vector $\boldsymbol{x} = [x(0), x(1), \dots x(N-1)]$

**Step 3.** The frequency interval is

$$\Delta\omega = 2\pi/(N\, \Delta t)$$

$$= \omega_0$$

Hence the spectrum will appear at intervals of the fundamental frequency, which is true for periodic signal, as is shown by the Fourier series expansion.[3]

The program can be written as follows:

```
% MATLAB Program to Compute FFT of a Periodic or
Nonperiodic Signal
X = fft(x)          ; calculates the FFT X(k) of the vector
                      x(n)
Xs = fftshift(X)    ; shifts the vector X(k) in symmetric
                      form
Xsm = abs(Xs)       ; magnitude spectrum
Xsp = angle(Xs)     ; phase spectrum
```

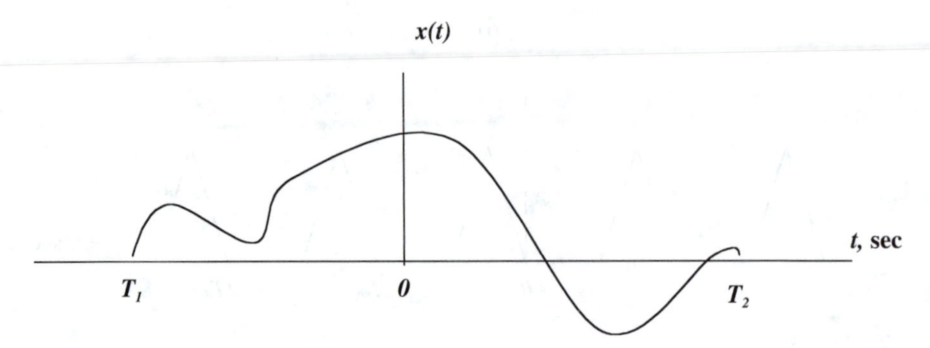


**FIGURE 3.5**
Nonperiodic signal $x(t)$.

*FFT evaluation of nonperiodic signals*

**Step 1.** Sample the signal $x(t)$, shown in Figure 3.5, *over the complete range* of the signal in the interval $T_1 \le t \le T_2$.. The sampling interval is:

$$\Delta t = (T_2 - T_1)/N$$

where $N$ is the number of points in the FFT.

**Step 2.** Generate the sampled signal $x(n)$, $n = 0, 1 \ldots N-1$. The input signal is stored as a vector $\boldsymbol{x} = [x(0), x(1), \ldots x(N-1)]$.

**Step 3.** The frequency interval is:

$$\Delta\omega = 2\pi/(N\,\Delta t)$$

The MATLAB program for FFT computation is identical to the one given in the previous section, for periodic signals.

## 3.2 Problem Solving

**Exercise 1: Solve the following problems, briefly outlining the important steps.**

a. Suppose that we are given an ideal low-pass discrete-time filter with frequency response:

$$H\left(e^{j\omega}\right) = 1, \qquad 0 \le |\omega| < \pi/4$$

$$= 0, \qquad \pi/4 < |\omega| \le \pi$$

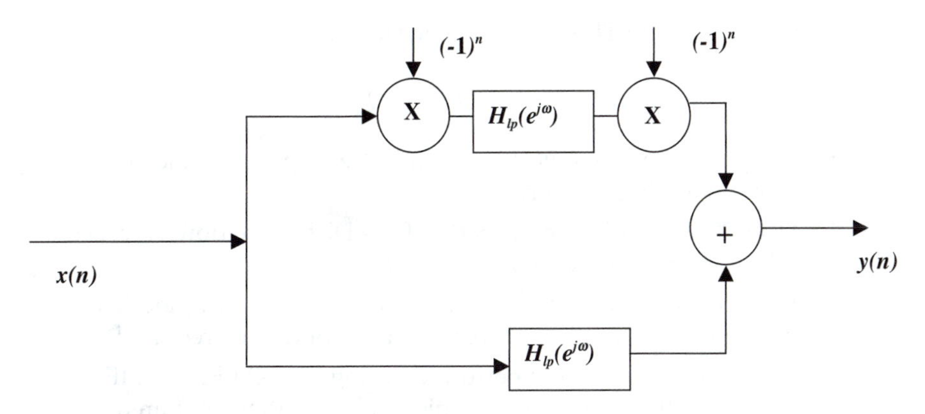


**FIGURE 3.6**
Figure for problem (b).

We wish to derive new filters from this prototype by manipulation of the impulse response $h(n)$.

i. Plot the frequency response $H_1(e^{j\omega})$ for the system whose impulse response is $h_1(n) = h(2n)$.

ii. Plot the frequency response $H_2(e^{jw})$ for the system whose impulse response is as follows:

$$h_2(n) = h(n/2),\ n = 0, \pm 2, \pm 4, \ldots$$

$$h_2(n) = 0, \qquad \text{otherwise}$$

iii. Plot the frequency response $H_3(e^{j\omega})$ for the system whose impulse response is $h_3(n) = e^{j\pi n}h(n)$.

b. Consider the system shown in Figure 3.6 with input $x(n)$ and output $y(n)$. The LTI systems shown with frequency response $H_{lp}(e^{jw})$ are ideal low-pass filters with cutoff frequency $\pi/4$ rad. and unity gain in the passband. Show that the overall system acts as an *ideal bandstop filter*, where the stopband is in the region $\pi/4 \le |\omega| \le 3\pi/4$.

c. Suppose we have two 4-point sequences $x(n)$ and $h(n)$ as follows:

$$x(n) = \cos(\pi n/2), \qquad n = 0, 1, 2, 3$$

$$h(n) = 2^n, \qquad n = 0, 1, 2, 3$$

i. Calculate the 4-point DFT $X(k)$.

ii. Calculate the 4-point DFT $H(k)$.

iii. Calculate $y(n) = x(n)$ ④ $h(n)$ by doing the circular convolution directly.

iv. Calculate $y(n)$ of part (iii) by multiplying the DFTs of $x(n)$ and $h(n)$ and performing an inverse DFT.

d. The output of an LTI discrete-time system is given by:

$$y(n) = x(n) * h(n)$$

where $x(n)$ is the input, $h(n)$ is the impulse response of the system, and $*$ denotes circular convolution.

i. Using the convolution property of the DFT, write down a procedure for obtaining $y(n)$, given $x(n)$ and $h(n)$.

ii. If the convolution were performed using $N$-point DFTs and IDFTs, determine the number of complex multiplications required.

iii. If the convolution were performed using *radix*-2 FFTs and IFFTs, determine the number of complex multiplications required.

iv. Compare the results of parts (ii) and (iii) for $N = 32$.

## 3.3 Computer Laboratory

**Exercise 2: Simulation of harmonic distortion in signal generators — Use of the FFT (Fast Fourier Transform)**

In this laboratory, the frequency spectra of periodic signals at the output of signal generator are studied analytically and by experiment. The periodic signals shown in Figure 3.7 are considered. There are several useful commands in MATLAB[4] to generate periodic signals, and some examples are given below.

***Periodic square pulse***

```
>> y = A*square(2*pi*f*t); generates a square wave
   vector y with peak amplitude A and frequency f Hz.
   The elements of y are calculated at the time
   instances of the vector t.
```

```
>> y = A*square(2*pi*f*t,duty); generates a square wave
   vector, with identical parameters as above, but with
   specified duty cycle. The duty cycle, duty, is the
   percent of the period in which the signal is
   positive.
```

***Aperiodic triangular pulse***

```
>> y = A*tripuls(t); generates samples of a continuous,
   aperiodic triangle at the points specified in array
```

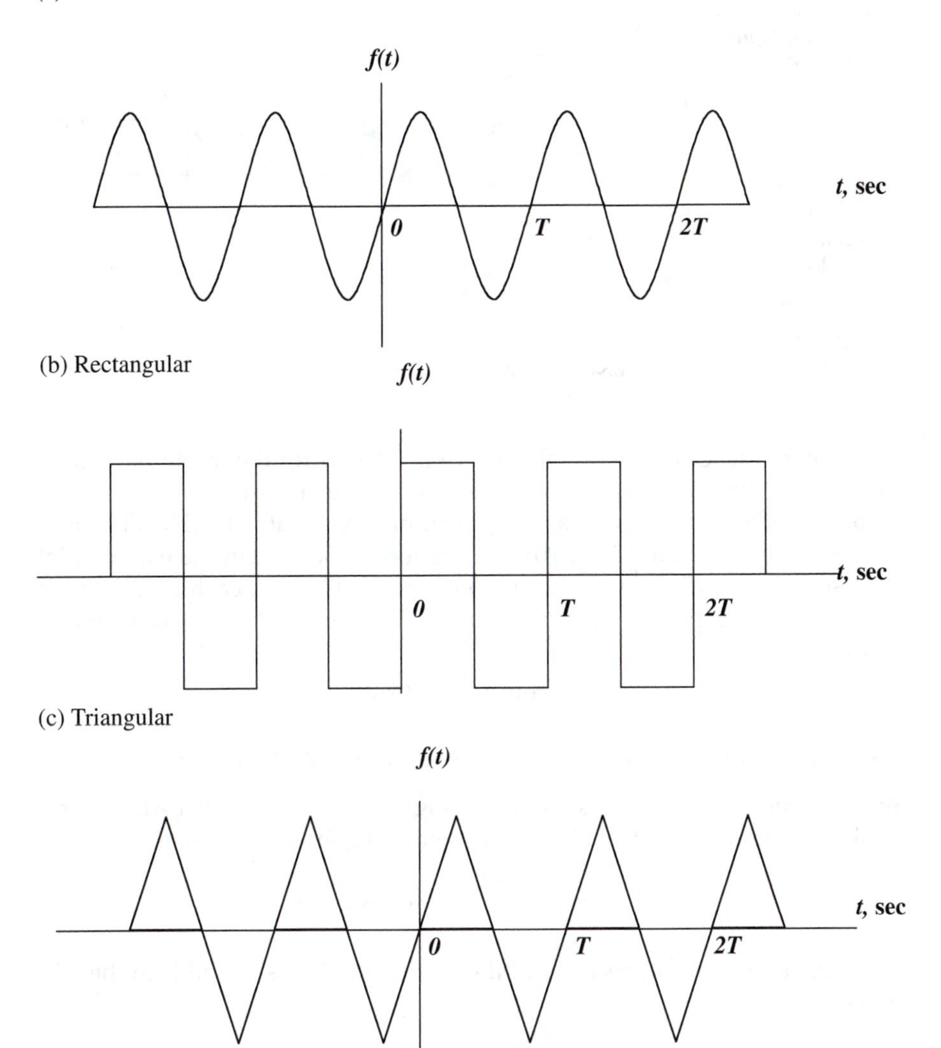


**FIGURE 3.7**
Periodic waveforms.

```
   t, centered about t = 0. By default, the triangle is
   symmetric and has duration of 1 sec.
>> y = tripuls(t,w); generates a triangle, with
   parameters as above, but duration of w, sec..
```

*Note:* Please also try other signal generation commands such as sin, cos, chirp, diric, gauspuls, pulstran, and rectpuls.

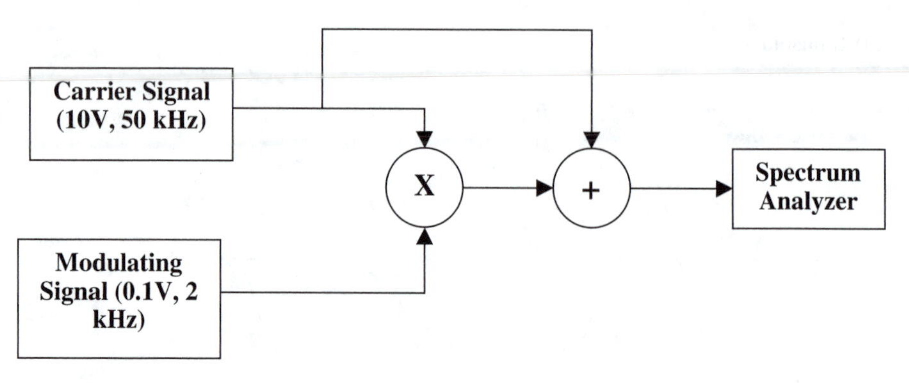


**FIGURE 3.8**
Model of Amplitude Modulation (AM) system.

Set the frequency of the signals to 1 MHz (11 kHz for the triangular wave) and the amplitude to 1 V. Compute the Fast Fourier Transform (FFT) of each of the periodic signals using an output resolution of $\Delta f = 1$ MHz. (The FFT can be implemented in MATLAB, see Section 3.1.3.) Obtain the exponential Fourier series coefficients $c_n$, $1 \le n \le 5$ (from the FFT) for each of the above waveforms. The power contained in the Fourier coefficients is given as follows:

$$Pc_n, \text{comp} = |c_n|^2 \text{ (mW)}$$

**Exercise 3: Simulation of Amplitude Modulation (AM) signals**

Simulate the AM system, as shown in Figure 3.8, using MATLAB or Simulink. The power contained in the Fourier coefficients is given as follows:

$$Pc_n, \text{comp} = |c_n|^2 \text{ (mW)}$$

Calculate the power spectrum of the carrier and two sidebands in the AM signal.

## 3.4 Hardware Laboratory

**Exercise 4: Measurement of harmonic distortion in signal generators**

Connect the output of the HP 3324A Synthesized Generator to the input of the HP 8590L Signal Analyzer as shown in Figure 1.8. Set the frequency of the generator to 1 MHz (11 kHz for the triangular wave) and the amplitude to 1 V. Measure the power spectrum (dBm) for each of the above signals to include the fundamental (1 MHz) and first four harmonics.

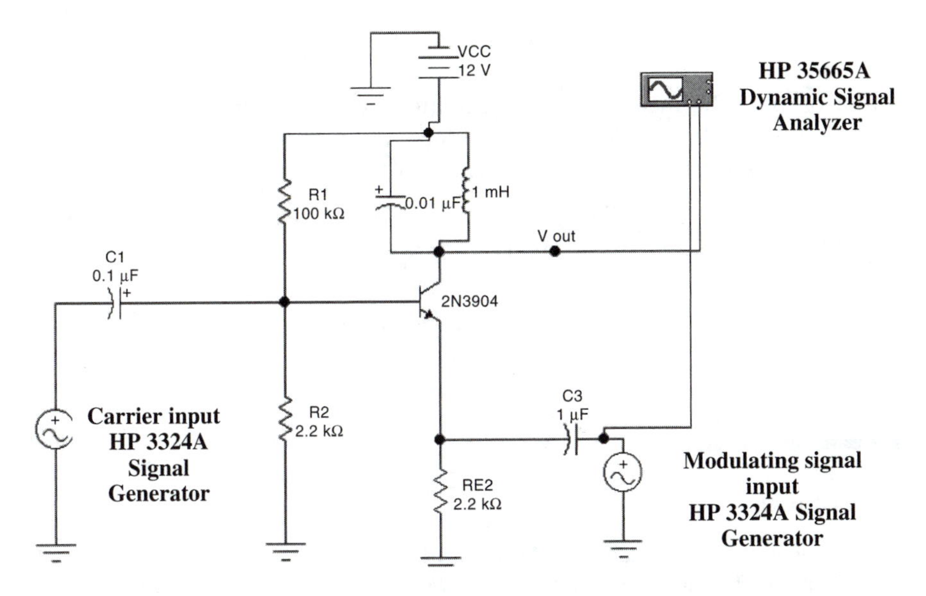


**FIGURE 3.9**
Practical Amplitude Modulation (AM) circuit.

a. Compare the measured and simulated power spectrum (from Section 3.3) of the fundamental and first four harmonics in mW, after normalizing the peak values of the fundamental to 1 mW (0 dBm).
b. Compute the percentage error between the computed and measured power spectrum (mW). The percentage error is defined as:

$$\% \text{ error} = \frac{|Pc_n, \text{comp} - Pc_n, \text{meas}|}{Pc_n, \text{comp}}$$

**Exercise 5: Measurement of spectrum in Amplitude Modulation (AM) signals.**

The circuit diagram for the Amplitude Modulation setup is shown in Figure 3.9. Connect one HP 3324A synthesized generator to the carrier input of the circuit. Set the carrier frequency at 50 kHz and amplitude at 10 volts. Similarly, connect another HP 3324A synthesized generator to the modulating signal input of the circuit. Set the modulating signal frequency at 2 KHz and the amplitude at 0.1 volt. Then connect the output of the AM circuit to channel 1 or channel 2 of the HP 35665A Dynamic Signal Analyzer, and observe the AM signal in both the time and frequency domains. Set a center frequency of 50 kHz and a span of 5 kHz on the Dynamic Signal Analyzer. Measure the power spectrum in dBm of the carrier and the two sidebands.

Obtain the power spectrum of the AM signal. The power contained in the Fourier coefficients is given as follows:

$$Pc_n, \text{ comp} = |c_n|^2 \text{ (mW)}$$

a. Compare the measured and simulated power spectrum (from Section 3.3) of the carrier and two sidebands in mW, after normalizing the peak values of the fundamental to 1 mW (0 dBm).

b. Compute the percentage error between the computed and measured power spectrum (mW). The percentage error is defined as:

$$\% \text{ error} = \frac{|Pc_n, \text{ comp} - Pc_n, \text{ meas}|}{Pc_n, \text{ comp}}$$

## References


1. Cooley, W. and Tukey, J.W. An Algorithm for the Machine Computation of Complex Fourier Series, *Mathematics of Computation*, 19, 297–301, 1965.
2. Cooley, J.W., Lewis, P.A.W., and Welch, P.D., Historical Notes on the Fast Fourier Transform, *IEEE Trans. Audio Electroacoustics*, Vol. AU-15, 76–79, 1967.
3. Lathi, B.P., *Linear Signals and Systems*, Oxford University Press, New York, 2001.
4. *Student Edition of MATLAB/Simulink*, Mathworks, Natick, MA.
5. Oppenheim, A.V and Schafer, R.W., with Buck, J.R., *Discrete-Time Signal Processing*, 2nd Edition, Prentice Hall, Upper Saddle River, NJ, 1998.
6. Kumar, B.P., *Digital Signal Processing Laboratory*, California State University, Sacramento, 2003.

# 4

# *Analog to Digital and Digital to Analog Conversion*

## 4.1 Brief Theory of A/D Conversion

Digital communications has proved to be a very efficient means of transporting speech, music, video, and data over different kinds of media. These media include satellite, microwave, fiber-optic, coaxial, and cellular channels. One special advantage that digital communication holds over analog communication is in the superior handling of noise in the channel.[1]

Baseband signals such as speech, music, and video are naturally occurring analog signals. Hence, the processes of analog to digital (A/D) conversion at the transmitter and digital to analog conversion (D/A) at the receiver are integral sections of the entire communication system shown earlier in Figure 1.2. We will now discuss *pulse code modulation* (*PCM*), which is one of the basic forms of A/D systems.

### 4.1.1 Pulse Code Modulation

Pulse code modulation was one of the earliest methods of A/D conversion.[2] The PCM process, as shown in Figure 4.1, converts an analog continuous-time signal, such as speech or music, into a digital binary bit stream. The three fundamental steps in the PCM process are time sampling, amplitude quantization, and binary encoding.

#### *4.1.1.1 Time Sampling*

The first step in the PCM process is *time sampling*, where the continuous-time signal $x(t)$, as shown in Figure 4.2a, is sampled uniformly at an interval of $T$ seconds. The output of the sampling process is the discrete-time signal $x(nT)$ or $x(n)$; $n = 0, 1, 2 \ldots N - 1$, as shown in Figure 4.2b.

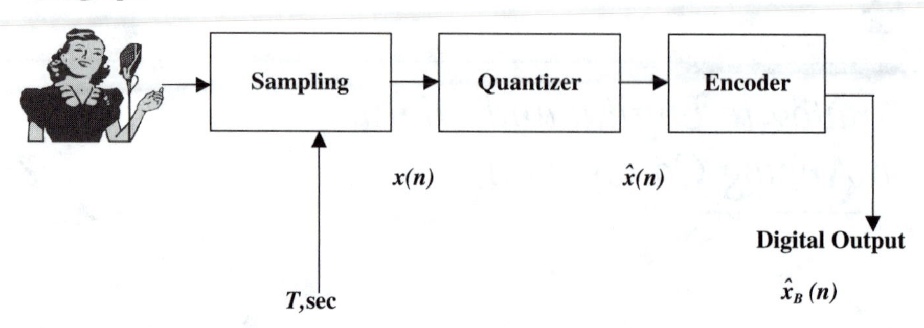


**FIGURE 4.1**
Pulse code modulation (PCM).

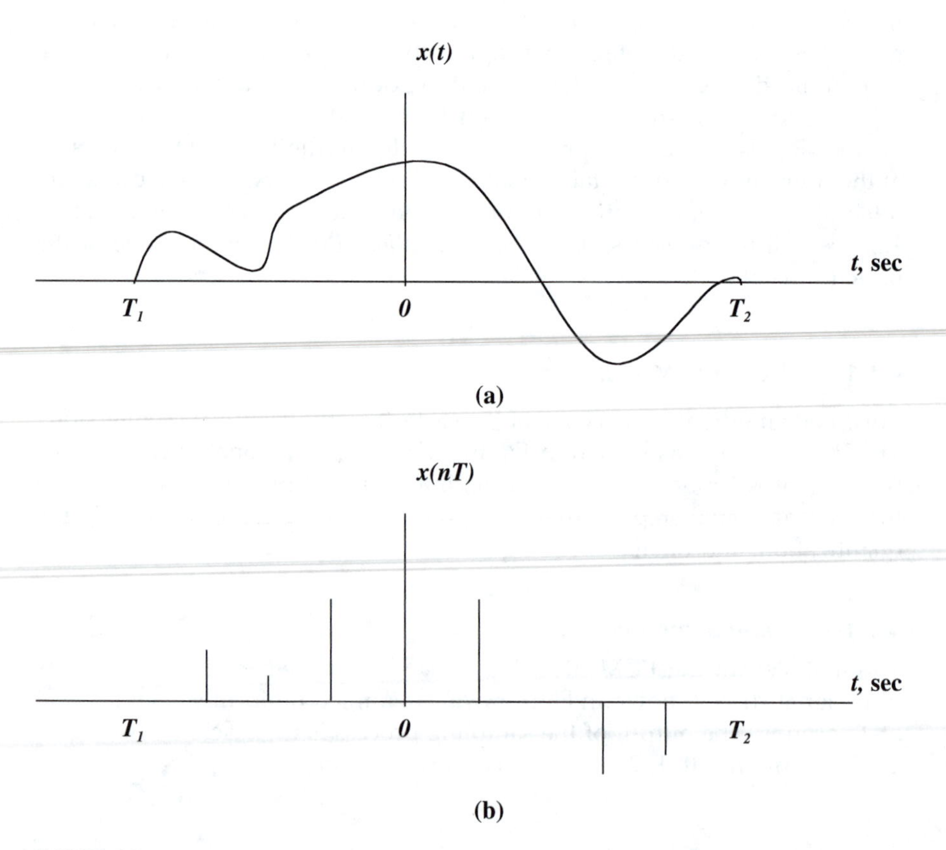


**FIGURE 4.2**
(a) Continuous-time signal $x(t)$, (b) uniformly sampled discrete-time signal $x(n)$.

Two important questions arise at this time:

- What is the appropriate value of the sampling interval, $T$ sec., or inversely, what is the appropriate value of the sampling frequency $f_s = 1/T$ in cycles per sec. or Hertz (Hz)?
- Is it possible to recover $x(t)$ exactly from the sample values $x(n)$: $n = 0, 1, 2 \ldots N-1$?

The answer to the first question is given by the Nyquist Sampling Theorem, which states: If $x(t)$ is a bandlimited signal with the maximum signal frequency $\Omega_m$, then $x(t)$ is uniquely determined from its samples $x(n)$: $n = 0, 1, 2 \ldots N-1$, if the sampling interval $T \leq \pi/\Omega_m$ seconds, or, alternately, if the sampling frequency $f_s \geq \Omega_m/\pi$ Hz.

The answer to the second question is given by the interpolation formula given below in Equation 4.1. If the sampling satisfies the Nyquist sampling theorem, then the recovered signal values (between the samples) is given by:

$$x_r(t) = \sum_{n=0}^{N-1} x(n) \frac{\sin\left[\pi(t-nT)/T\right]}{\pi(t-nT)/T} \tag{4.1}$$

However, *practical sampling*, which will be studied in Section 4.2, is different than the ideal sampling described in this section and by Equation 4.1. One of the practical problems in ideal sampling is the impossibility of generating ideal impulses with zero time width.

#### 4.1.1.2 Amplitude Quantization

The second stage in the A/D process is amplitude quantization, where the sampled discrete-time signal $x(n)$, $n=0, 1, 2 \ldots N-1$ is quantized into a finite set of output levels $\hat{x}(n)$, $n = 0, 1, 2 \ldots N-1$. The quantized signal $\hat{x}(n)$ can take only one of $L$ levels, which are designed to cover the dynamic range $-x_M \leq x(n) \leq x_M$, where $x_M$ is the maximum amplitude of the signal. Both uniform and nonuniform quantizers will be considered in this section.

##### Uniform Quantizer

The design of an *L-level uniform quantizer* is detailed below in a 4-step process.

**Step 1: Dynamic range of the signal**

Fix the dynamic range of the sampled signal $-x_M \leq x(n) \leq x_M$.

**Step 2: Step size of quantizer**

The step size of the uniform quantizer is given as:

$$\text{Step size } \Delta = \frac{2x_M}{L} \tag{4.2}$$

The step size can be either integer or fraction and is determined by the number of levels $L$. For binary coding, *$L$ is usually a power of* 2, and practical values are 256 (=$2^8$) or greater.

**Step 3: Quantizer implementation**

Draw the input-output or staircase diagram of the quantizer, as shown in Figure 4.3. The x-axis of the staircase diagram represents the input sampled signal x (n), and the y-axis represents the quantized output $\hat{x}(n)$.

As is seen from Figure 4.3a, the input levels are in integral multiples of $\Delta/2$, while the output levels are in integral multiples of $\Delta$, *with output zero level included*. Such a quantizer is termed a *mid-tread* quantizer, whereas a *mid-riser* quantizer, as shown in Figure 4.3b, does not include output zero level and has the reverse structure of the mid-tread quantizer.

**Step 4: Quantizer error and SNR**

The quantizer error is calculated as

$$e(n) = \hat{x}(n) - x(n),\ n = 0, 1, 2 \ldots N-1$$

A figure of merit of the quantizer is defined by the *Quantizer Signal-to-Noise ratio* (*SNR*) given as:

$$SNR = 10\log\left[\frac{\sigma_x^{\,2}}{\sigma_e^{\,2}}\right] \tag{4.3}$$

In Equation 4.3, the *variance of the input signal* $x(n)$ is given as

$$\sigma_x^{\,2} = \overline{x^2(n)} - [\overline{x(n)}]^2$$

where $\overline{x^2(n)}$, the *mse* (*mean squared value*) of the input signal is given by

$$\overline{x^2(n)} = \frac{x^2(0) + x^2(1) + x^2(2)\ \ldots\ x^2(N-1)}{N} \tag{4.4}$$

and, the mean value of the input signal, $\overline{x(n)}$, is given by

$$\overline{x(n)} = \frac{x(0) + x(1) + x(2)\ \ldots\ x(N-1)}{N} \tag{4.5}$$

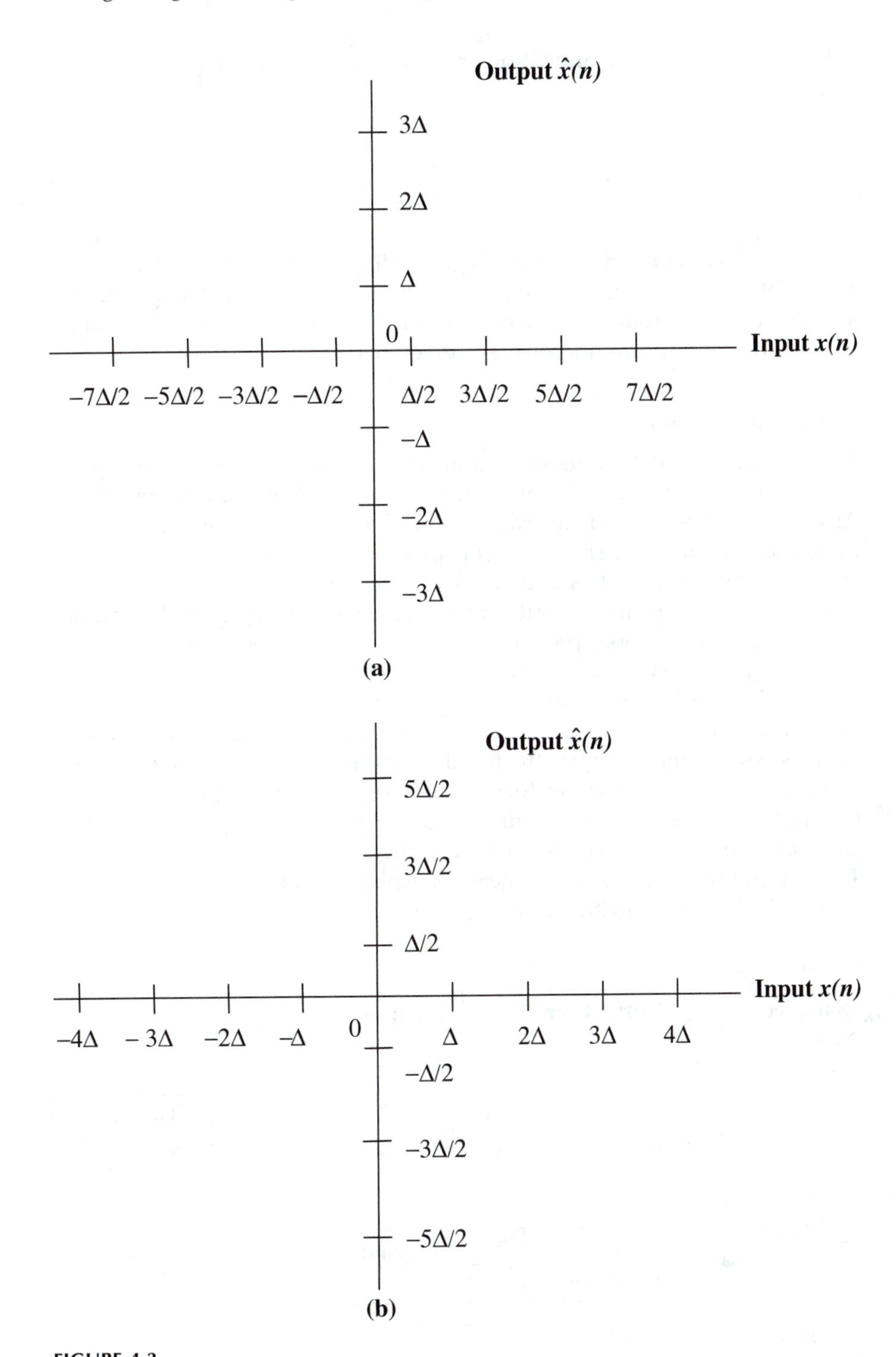


**FIGURE 4.3**
Input-output staircase diagram of uniform quantizer. (a) Mid-tread quantizer, (b) Midriser quantizer.

The variance of the quantization error is given by a simplified expression below:

$$\sigma_e^2 = \frac{\Delta^2}{12} \tag{4.6}$$

*Practical quantizers* are used for high-quality music work at *SNR* values around 90 dB. *Nonuniform quantizers,* such as μ-law and *A*-law quantizers, are widely used around the world to improve the *SNR* value. A detailed analysis of nonuniform quantizers is given below.

### Nonuniform Quantizer

The most important nonuniform quantization technique is logarithmic quantization (μ-*law* in the U.S., Japan, and Canada and *A-law* in Europe, Africa, Asia, South America, and Australia), which has been used very successfully for speech digitization. This technique evolved from the fundamental property of speech, which has a gamma or Laplacian probability density in amplitude, highly peaked about zero value. Hence, even though low amplitudes of speech are more probable than large amplitudes, a uniform quantizer amplifies all signals equally.

The principle behind nonuniform quantization is to pre-process (compress) the sampled signal before it enters the uniform quantizer, such that the processed signal occupies the full dynamic range of the quantizer. However, the output of the uniform quantizer has to be post-processed (expanded) to extract the true quantized signal. The dual process is called *logarithmic companding,* which is a combination of *compression* and *expanding*. The nonuniform quantization process is explained below in a series of four steps, which is also illustrated in Figure 4.4.

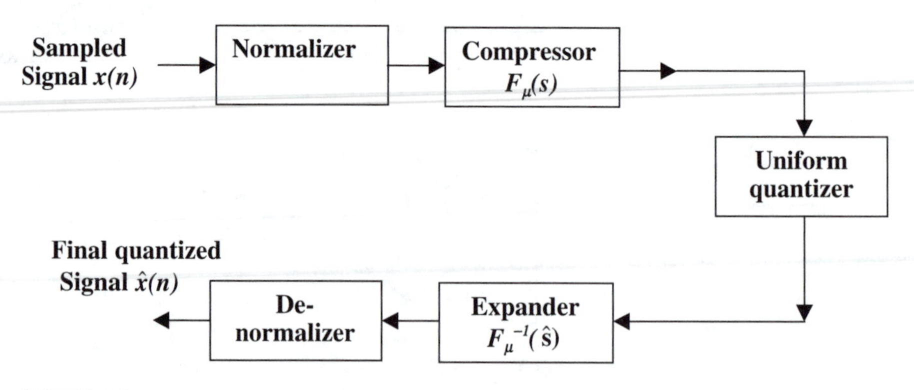


**FIGURE 4.4**
Nonuniform (μ-law) quantization process.

**Step 1: Dynamic range and normalizing the sampled signal**

Fix the dynamic range of the sampled signal $-x_M \le x(n) \le x_M$. Normalize the sampled signal $x(n)$ by its peak amplitude $x_M$, to yield the normalized signal $s(n)$:

$$s(n) = \frac{x(n)}{x_M}$$

with a dynamic range $-1 \le s(n) \le 1$.

**Step 2: Signal compression**

This step basically pre-processes the sampled signal, to provide more amplification to the lower amplitude samples and less amplification to the higher amplitude samples. The compression function is given below:

$$F_\mu(s) = \frac{\ln\left(1+\mu|s|\right)}{\ln(1+\mu)}\operatorname{sgn}(s) \tag{4.7}$$

where $s$ is the normalized sampled signal and $\mu$ is the compression parameter, usually taken as 255.

**Step 3: Processing by uniform quantizer**

The compressed output, $F_\mu(s)$, is input to a uniform $L$-level quantizer, which has been described in detail in the previous section.

**Step 4: Signal expanding**

The output of the uniform quantizer, $\hat{s}(n)$, is passed through the inverse expanding function, in order to re-synthesize the input signal. The expanding function is given below:

$$F_\mu^{-1}(\hat{s}) = \frac{1}{\mu}\left[(1+\mu)^{|\hat{s}|} - 1\right]\operatorname{sgn}(\hat{s}) \tag{4.8}$$

where $-1 \le \hat{s}(n) \le 1$.

**Step 5: Signal de-normalization**

The final step in the nonuniform quantization process is the de-normalization of the signal, $\hat{s}(n)$, to yield the final quantized signal:

$$\hat{x}(n) = \hat{s}(n)x_M$$

#### 4.1.1.3 Binary Encoding

The third and last stage in the A/D process is binary encoding, where the quantized signal $\hat{x}(n)$, $n = 0,1,2 \ldots N-1$ is encoded to yield the final *digital signal* $\hat{x}_B(n)$, $n = 0,1,2 \ldots N-1$. As an illustration, if the number of quantizer levels $L = 8 = 2^3$, the number of binary bits required to encode all the $L$ levels is 3. Table 4.1 illustrates the encoding procedure, using *two's complement coding*, which is very convenient in the decoding process at the receiver.

The two's complement code (TCC) is quite easily obtained from the offset binary code (OBC), by *complementing the left-most bit of the OBC*. Some of the advantages of the two's complement code are as follows:

- The decimal form of the TCC includes *both positive and negative numbers* and is given by the following equation:

$$\text{Decimal number} = -a_0 2^0 + a_1 2^{-1}\, a_2 2^{-2} + \ldots\, a_B 2^{-B} \tag{4.9}$$

  where the original binary number is $[a_0\ a_1\ a_2\ a_B \ldots a_B]$.
- The decoding process at the receiver is more efficient, as illustrated in Table 4.2.

**TABLE 4.1**
Binary Encoding Process

| Quantizer Level (for $L$= 8) | Offset Binary Code (3-Bit) | Two's Complement Code (3-Bit) |
|---|---|---|
| –4 | 000 | 100 |
| –3 | 001 | 101 |
| –2 | 010 | 110 |
| –1 | 011 | 111 |
| 0 | 100 | 000 |
| 1 | 101 | 001 |
| 2 | 110 | 010 |
| 3 | 111 | 011 |

**TABLE 4.2**
Binary Decoding Process

| Two's Complement Code (3-Bit) | Decimal Value (from Equation 4.9) | Actual Quantizer Level |
|---|---|---|
| 100 | –1 | –4 |
| 101 | –3/4 | –3 |
| 110 | –1/2 | –2 |
| 111 | –1/4 | –1 |
| 000 | 0 | 0 |
| 001 | 1/4 | 1 |
| 010 | 1/2 | 2 |
| 011 | 3/4 | 3 |

From Table 4.2, it is seen that the recovered quantized value from the binary bit stream can be obtained easily as follows:

$$\text{Quantized value} = \text{Decimal value of TCC} \times \text{Peak value of sample value } (x_M)$$

## 4.2 Problem Solving

**Exercise 1: Solve the following problems, briefly outlining the important steps.**

a. Design a uniform 8-level quantizer designed for an input signal with a dynamic range of ± 10 volts.
   i. Calculate the quantization error vector for an input signal of $x(n) = [-4.8\ \ -2.4\ \ 2.4\ \ 4.8]$.
   ii. Calculate the quantization error for the same input signal if the quantizer is preceded by a $\mu = 255$ compander (compressor/expander).

b. A continuous signal $x_c(t)$ has a Fourier transform $X_c(j\Omega)$, which exists in the range $\Omega_0/2 \leq |\Omega| \leq \Omega_0$, and is zero elsewhere in the frequency. This signal is sampled with sampling period $T = 2\pi/\Omega_0$ to form the discrete-time sequence $x(n) = x_c(nT)$.
   i. Sketch the Fourier transform $X(e^{j\omega})$ for $|\omega| < \pi$.
   ii. The signal $x(n)$ is to be transmitted across a digital channel. At the receiver, the original signal $x_c(t)$ must be recovered. Draw a block diagram of the recovery system and specify its characteristics. Assume that ideal filters are available.
   iii. In terms of $\Omega_0$, for what range of values of $T$ can $x_c(t)$ be recovered from $x(n)$?

c. A TV signal has a bandwidth of 4.5 MHz. This signal is sampled, quantized, and binary coded to obtain a PCM signal.
   i. Determine the sampling rate if the signal is to be sampled at a rate 20% above the Nyquist rate.
   ii. If the samples are quantized into 1024 levels, determine the number of binary pulses q required to encode each sample.
   iii. Determine the binary pulse rate (pulses/sec or bits/sec) of the binary coded signal.

## 4.3 Computer Laboratory

**Exercise 2: Simulation of A/D Sample and Hold (S & H) circuits using Simulink[3]**

A Sample and Hold (S & H) circuit is the key element required in the conversion of a voltage from analog to digital form. The S & H circuit samples the input analog voltage periodically and then holds it constant. The circuit following the S & H circuit is the quantizer circuit, which converts the sampled signal into the digital signal.

In this laboratory, the aim is to study the effects of practical sampling and quantization on the input signal and also on the reconstructed signal at the receiver.

***Practical circuit for A/D conversion:***

The schematic of a practical A/D circuit is shown in Figure 4.5a. The ideal S & H, shown in the figure, is equivalent to impulse train modulation

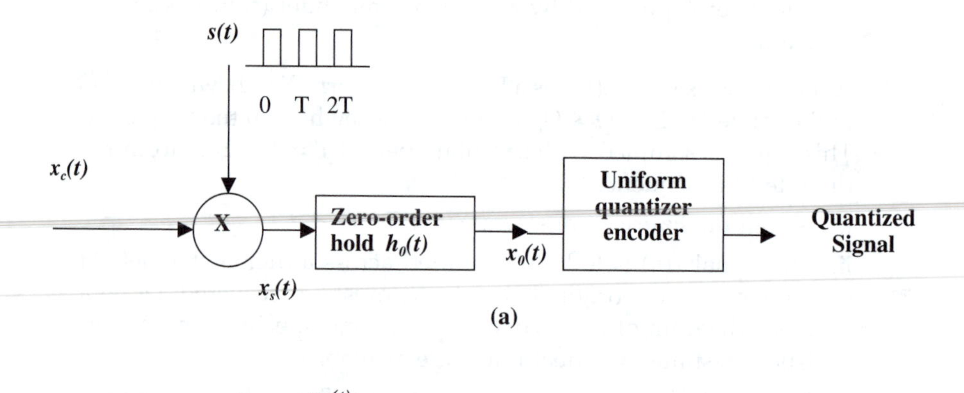


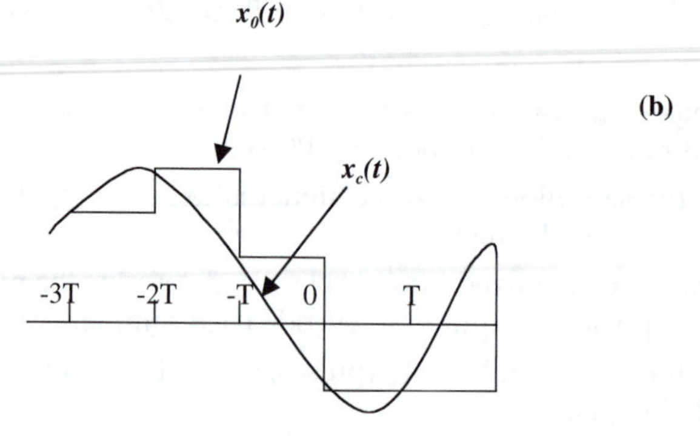


**FIGURE 4.5**
Practical sampling using sample and hold (S & H) circuit.

followed by linear filtering with the Zero-Order Hold (ZOH) system. The output of the ZOH system is the staircase waveform shown in Figure 4.5b. The sample values are held constant during the sampling period of $T$ seconds.

In the design of the sample and hold circuit on Simulink, the following three important blocks will have to be designed accurately:

- *The source signal block $x_s(t)$:* Assume a sinusoidal signal having a peak-to-peak amplitude of 1 V, and a frequency of 20 Hz. Since this block's parameters are fixed, no further design is necessary on this block.
- *The pulse train block $s(t)$:* Two important parameters will have to be designed for this block. The first is the pulse amplitude, and the second is pulse period $T$ sec. You can assume a rectangular pulse with 50% duty cycle (i.e., half period on and half period off). Because this pulse train samples the source signal, its frequency should be many times higher than that of the source signal.
- *The Zero-Order Hold block:* One important parameter will have to be designed for this block, which is the sampling period of the hold circuit. The sampling period of the circuit should be sufficient to hold the sample value over each period of the pulse train.

***Practical circuit for D/A conversion:***

The schematic of a practical D/A circuit is shown in Figure 4.6

a. Select an appropriate audio signal from the Simulink DSP blockset library as the test signal in this simulation. Plot the signal on the scope and the FFT scope to obtain the frequency content of the signal. This will provide information on the maximum frequency content of the signal and the required sampling rate limits. Sample the signal at the Nyquist rate.
b. Design the required parameters of the A/D circuit given in Figure 4.5a to obtain the appropriate staircase pattern as shown in Figure 4.5b. Plot the output of the Zero-Order Hold circuit as seen on the scope block of the Simulink program.
c. Design a uniform quantizer to convert the sampled signal into quantized signal output in numerical or binary form.

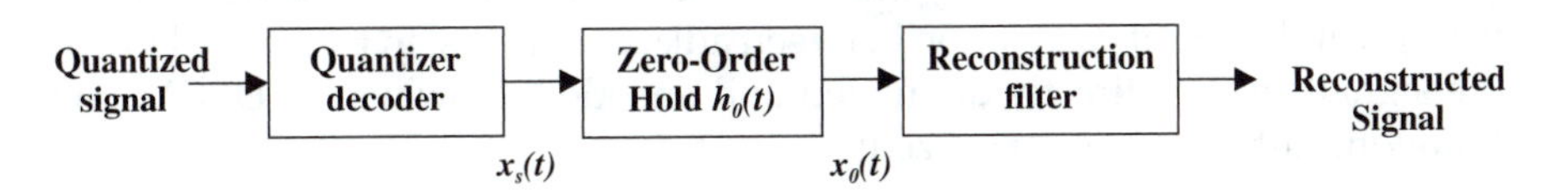


**FIGURE 4.6**
Schematic of digital to analog (D/A) conversion circuit.

d. Design the required parameters of the D/A circuit given in Figure 4.6 to reconstruct the signal at the receiver. The reconstruction filter is modeled as a low-pass filter (analog or digital) with cutoff frequency as the sampling frequency utilized in the A/D process.
e. Plot the reconstructed signal, compared with transmitted analog signal, and plot the error signal.
f. Repeat the entire simulation for a case of undersampling: choose a sampling frequency smaller than the Nyquist rate (e.g., half the Nyquist rate), and plot the transmitted signal, reconstructed signal, and the error signal.
g. Repeat the entire simulation for a case of oversampling: choose a sampling frequency larger than the Nyquist rate (e.g., twice the Nyquist rate), and plot the transmitted signal, reconstructed signal, and the error signal.

**Exercise 3: Simulation of A/D sample and hold (S & H) circuits with nonuniform quantization**

Repeat Exercise 2, steps a through c, however, with the following modifications:

- Introduce a μ-*law compressor* before the uniform quantizer, as shown in Figure 4.4. Similarly introduce a μ-*law expander* after the uniform quantizer. Assume $\mu = 255$.
- As in Exercise 2, repeat steps d through f, and plot the transmitted signal, reconstructed signal, and the error signal for the cases of undersampling, oversampling, and Nyquist sampling.
- Compare the error in reconstruction, between the cases of uniform quantization and nonuniform quantization.

**Exercise 4: Simulation of Differential Pulse Code Modulation (DPCM) system**

Simulate the DPCM system, shown in Figure 4.7, using Simulink.

*a. Transmitter*

Assume an input signal: $s(t) = 10\,sin(5\pi t) + 5\,sin(8\pi t)$. In this simulation, one period or multiple periods of the signal can be processed. The input analog signal $s(t)$ is sampled at a rate much higher than the Nyquist rate (~25 to 50 times). This generates very closely spaced samples $s(nT)$, which have a very great degree of correlation between adjacent values. In traditional PCM, the signal $s(nT)$ is directly quantized and encoded. However in DPCM, the following difference is quantized:

$$e(t) = s(nT) - s(\overline{n-1}T)$$

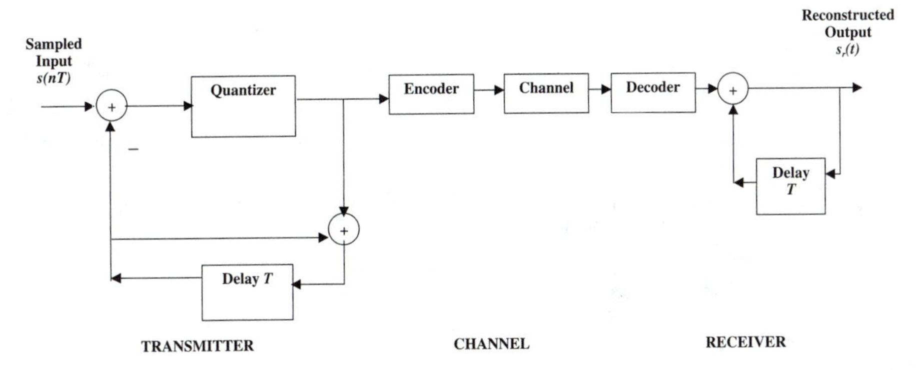


**FIGURE 4.7**
Block diagram of Differential PCM transmitter and receiver.

The difference signal is quantized as follows:

$$\hat{e}(t) = \delta \text{ if } e(t) > 0$$

$$\hat{e}(t) = -\delta \text{ if } e(t) < 0$$

$$\hat{e}(t) = 0 \text{ if } e(t) = 0$$

or compactly, $\hat{e}(t) = sgn[e(t)]$, where sgn is the signum function and the step size should be selected to satisfy the condition: $\delta << |s(t)|_{\max}$. Thus the final signal $\hat{e}(t)$ consists of pulses with amplitude $\pm\delta$. Plot the quantized signal $\hat{e}(t)$ for at least one period of the original signal $s(t)$.

*b. Channel*

Model the channel as a system gain of 1.0.

*c. Receiver*

The receiver consists of an integrator, which sums the pulses $e(n)$, and generates the reconstructed signal $s_r(t)$.

i. Plot the input and reconstructed signals on the same graph, and determine the mean-squared error between them.
ii. Plot the error signal between the input and the reconstructed signals.

## 4.4 Hardware Laboratory

### Exercise 5: Design and construction of a simple Sample and Hold (S & H) circuit

The S & H circuit using an FET switch, shown in Figure 4.8,[4] can sample rapidly changing voltages that arise from the input signal $x_c(t)$. The op-amp acts as a high input-impedance voltage follower.

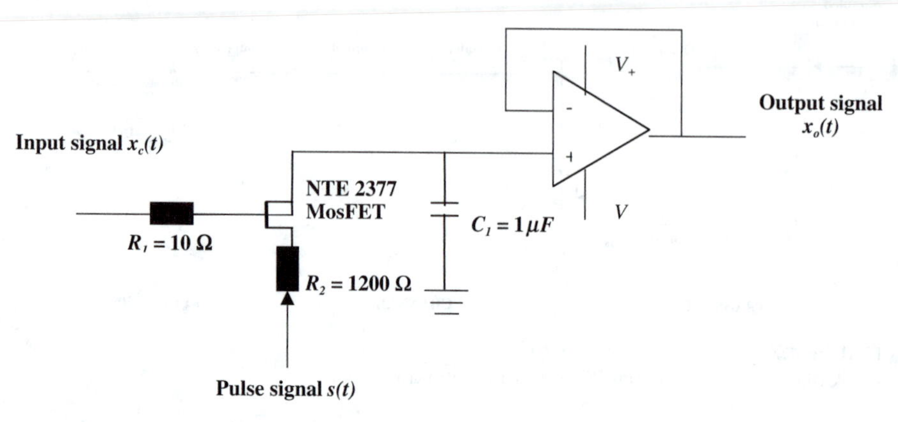


**FIGURE 4.8**
Practical MosFET sample and hold circuit.

When a pulse train $s(t)$ is high at the sample input, the FET is turned on (during the *on* cycle) and acts as low resistance to the input signal. When the sample pulse is absent, the MOSFET is turned off and acts as high impedance. The desired voltage is held by capacitor $C_1$, which is isolated from the output by the high input impedance op-amp. When the switch is closed, the capacitor charges to $x_{c(\max)}$. After the switch is opened, the capacitor remains charged and $x_o(t)$ will be at the same potential as the capacitor. The sampled voltage will be held temporarily, the time being determined by leakage in the circuit.

a. Connect the circuit as shown in Figure 4.9, and apply a 1 kHz (input signal frequency) sinusoidal signal to the input of the S & H circuit. Use a 10 kHz (sampling frequency) pulse signal to drive the sample input of the S & H circuit. Observe the sampled output at the output of the circuit on an oscilloscope.
b. Repeat the experiment for the maximum input signal frequency possible. Please note that the sample frequency should be accordingly increased in order to obtain the required number of samples.
c. Plot the spectra of the input and output signals of the S & H circuit on the HP35665A Dynamic Signal Analyzer. Comment on the differences between the two spectra.

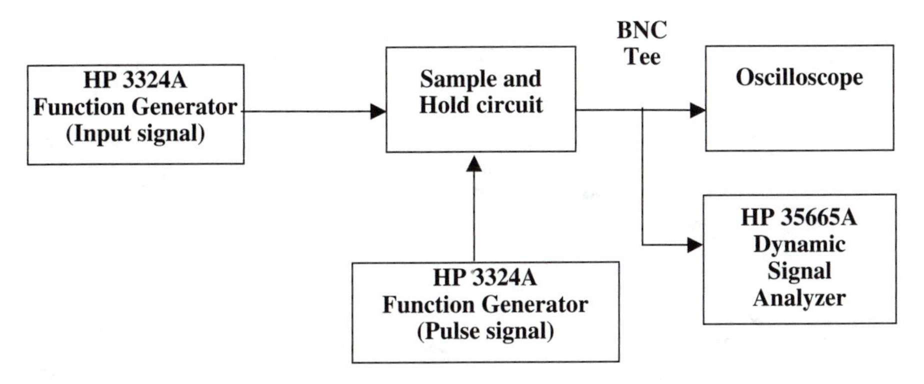


**FIGURE 4.9**
Measurement setup for sample and hold circuit.

## References

1. Lathi, B.P., *Modern Digital and Analog Communication Systems*, 3rd Edition, Oxford University Press, New York, 1998.
2. Gibson, J.D., *Principles of Digital and Analog Communications*, Second Edition, Prentice Hall, MA, 1993.
3. *Student Edition of MATLAB/Simulink*, Mathworks, Natick, MA. Version 5.3, 1999.
4. Oppenheim, A.V and Schafer, R.W., with Buck, J.R., *Discrete-Time Signal Processing*, 2nd Edition, Prentice Hall, Upper Saddle River, NJ, 1998.
5. Kumar, B.P., *Digital Signal Processing Laboratory*, California State University, Sacramento, 2003.

# 5

# *Digital Filter Design I: Theory and Software Tools*

## 5.1 Brief Theory of Digital Filter Design

*Signal filtering* is one of the most important operations in many electrical engineering systems.[1] However, the most widely used applications of filtering are found in communications engineering, including:

- *Frequency selection process*: In the car radio, for example, a tunable bandpass filter enables us to select our favorite AM or FM radio channel. In a more recent application, a mobile phone switches its carrier frequency rapidly when it moves around from cell to cell.
- *Signal demodulation*: In amplitude or frequency modulation, low-pass filters are used to filter out the low-frequency baseband signal from the high-frequency modulated signal.
- *Removal of signals from noise*: Generic filters such as low-pass or bandpass and specific filters such as *Wiener* and *Matched filters*[1] are used to extract audio and video information from noisy signals. An audio signal can be represented as a one-dimensional time-dependent function $x(t)$, representing signal amplitude, in volts, whereas video signals are two-dimensional functions of space $f(x,y)$, representing the image intensity. A more detailed discussion on two-dimensional video analysis and filtering will be provided in Chapter 6.
- *Analysis of practical signals*: Biomedical signals such as the EKG (heart) and EEG (brain) provide valuable information into the workings of specific areas in the human body.[2] Filters are essential to remove noise, or "smoothen" the received biomedical signal, before the signal is analyzed, using tools such as the FFT.

### 5.1.1 Analog and Digital Filters

Once a desired filter response $H(j\Omega)$ is specified, for example, the lowpass filter response in Figure 5.1a, then the filter can be realized in *analog form* as

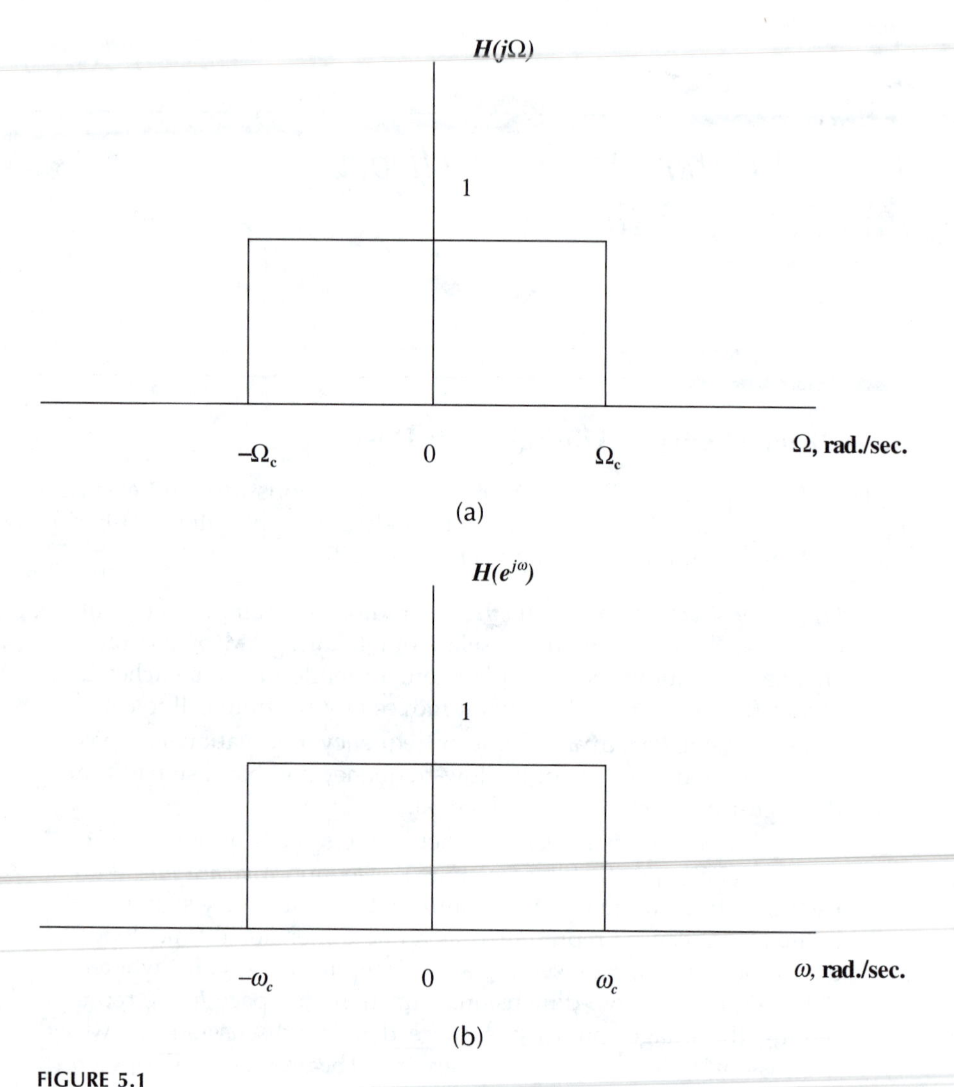


**FIGURE 5.1**
(a) Analog frequency response of ideal low-pass filter; (b) digital frequency response of ideal low-pass filter.

shown in Figure 5.2. The component resistor and capacitor values are designed using Equation 5.1.[3]

$$\text{Cutoff frequency } f_c(Hz) = \frac{\Omega_c}{2\pi} = \frac{1}{2\pi RC} \tag{5.1}$$

where the resistor value is in ohms and capacitor value is in Farads.

However, the same filter response, as shown in Figure 5.1a can also be realized using a *digital filter,* as shown by the system block diagram in

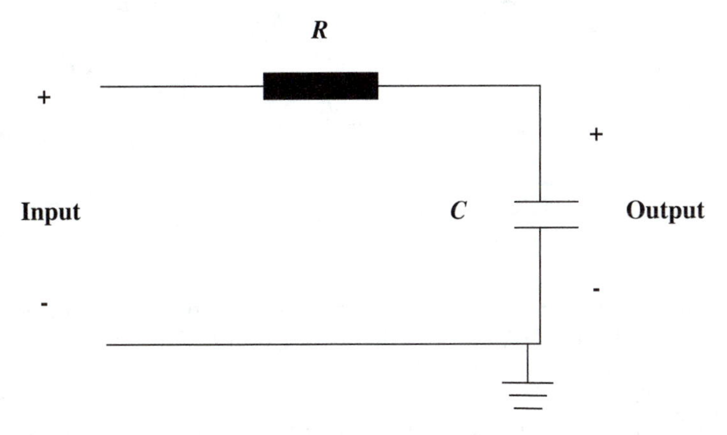


FIGURE 5.2
Analog realization of low-pass filter.

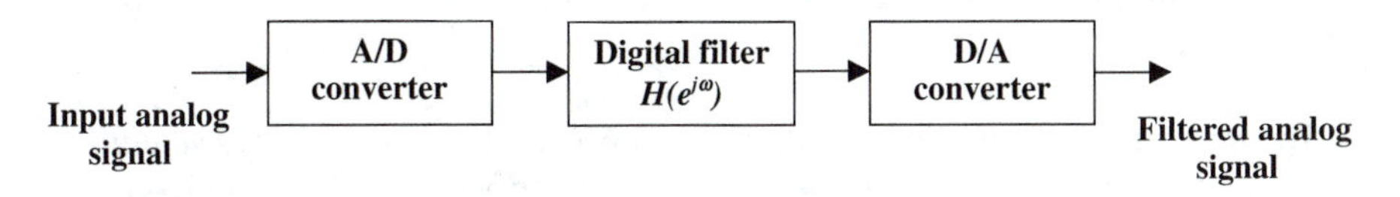


FIGURE 5.3
Digital realization of low-pass filter.

Figure 5.3. The analog signal $x(t)$ is converted into a discrete-time signal $x(n)$, which is processed by the digital filter, to yield a discrete-time output $y(n)$. Finally, the discrete-time output $y(n)$ is converted to its analog form $y(t)$. The *cutoff frequency of digital filter* response $H(e^{j\omega})$, as shown in Figure 5.1b, is related to the *analog cutoff frequency* through the important analog-digital frequency relation:

$$\omega = \Omega T \tag{5.2}$$

where $T$ (sec.) is the sampling interval of the discrete-time system.

Hence, the unit of analog frequency, $\Omega$, is radians/sec, while the unit of digital frequency, $\omega$, is radians.

The digital filter, shown in Figure. 5.2, can be realized using a Digital Signal Processor (DSP). The DSP can be programmed to act as any kind of filter. This is one of the main advantages of digital systems.

### 5.1.2 Design Techniques for FIR and IIR Digital Filters

As with any other discrete-time system, any digital filter can be described by the linear constant-coefficient difference equation of the form:

$$a_0 y(n) + a_1 y(n-1) + \ldots\ a_N y(n-N) = b_0 x(n) + b_1 x(n-1) + \ldots\ b_M x(n-M)$$

or, rewriting in terms of the current output $y(n)$:

$$y(n) = -\frac{a_1}{a_0} y(n-1) - \ldots - \frac{a_N}{a_0} y(n-N) + \frac{b_0}{a_0} x(n) + \frac{b_1}{a_0} x(n-1) + \ldots \frac{b_M}{a_0} x(n-M) \tag{5.3}$$

The difference equation above describes a digital filter of order $N$, where, in general, $N > M$. The two important classifications of digital filters are *finite impulse response* (*FIR*) and *infinite impulse response* (*IIR*).

- FIR filters have all the feedback coefficients $[a_1, a_2, \ldots a_N]$ equal to zero.
- IIR filters are characterized by at least one nonzero element in the vector $[a_1, a_2, \ldots a_N]$.

Design methods of digital filters can be broadly divided into *analytical methods* and *computer aided methods*. Most filter designs can be realized efficiently by applying software such as MATLAB, but some analytical techniques are also significant.

#### 5.1.2.1 Analytical Techniques for IIR Digital Filter Design

IIR digital filter techniques are essentially based on transformation of efficient analog filters, such as Butterworth and Chebyshev filters, into corresponding digital filters. This can be achieved either in the time domain $h_c(t) \to h(n)$, or in the frequency domain $H_c(j\Omega) \to H(e^{j\omega})$, or in the complex frequency domain $H_c(s) \to H(z)$. The two main types of analog-to-digital filter transformation techniques are *impulse invariance* and *bilinear transformation*, which are explained below.

##### Impulse invariance method

In the impulse invariance method, the impulse response of the digital filter, $h(n)$, is directly proportional to the uniformly sampled version of the corresponding analog filter $h_c(t)$, i.e.,

$$h(n) = T\, h_c(t)/_{t=nT} = T\, h_c(nT) \tag{5.4}$$

where $T$ represents the sampling interval. The corresponding transformation in the frequency domain can be derived as follows:

$$H\left(e^{j\omega}\right)=H_c\left(j\frac{\Omega}{T}\right) \tag{5.5}$$

through the frequency transformation given in Equation 5.2.

***Bilinear transformation method***

The bilinear transformation of the analog filter system function, $H_c(s)$, yields the corresponding digital filter system function, $H(z)$, and is obtained through the transformation:

$$s=\frac{2}{T}\left(\frac{1-z^{-1}}{1+z^{-1}}\right) \tag{5.6}$$

Substituting $s = \sigma + j\omega$ and $z = e^{j\omega}$ into Equation 5.6, and equating real and imaginary parts of the resulting equation, the following frequency transformation is obtained:

$$\Omega=\frac{2}{T}\tan\left(\omega/2\right) \tag{5.7}$$

#### *5.1.2.2 Analytical Techniques for FIR Filter Design*

One of the most widely used methods of FIR digital filter design is the *window method*, which will be briefly explained below, in a series of steps.

**Step 1: Specification of the desired filter response $H_d(e^{j\omega})$**

For example, a desired low-pass response is shown in Figure 5.4.

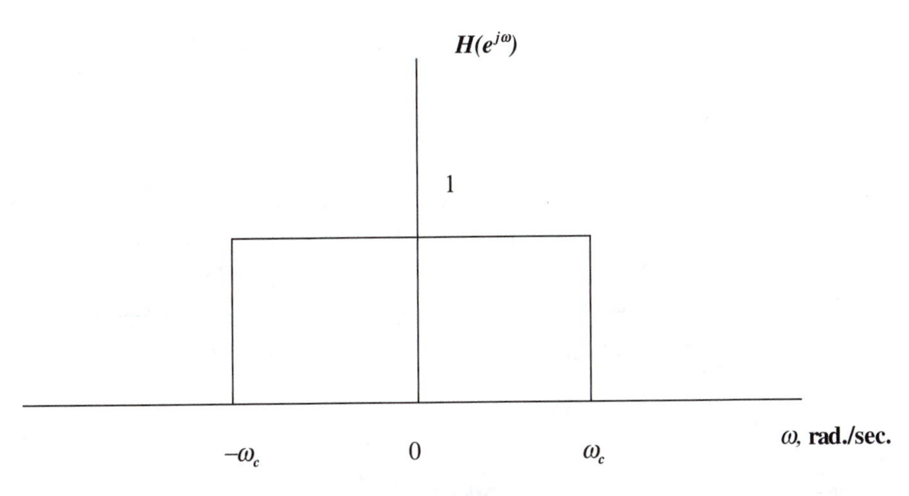


**FIGURE 5.4**
Desired low-pass response of digital filter.

**Step 2: Obtaining the ideal filter impulse response $h_d(n)$**

The impulse response can be obtained by the inverse DTFT, from Chapter 2, as follows:

$$h_d(n) = \frac{1}{2\pi} \int_{-\pi}^{\pi} H_d\left(e^{j\omega}\right) e^{j\omega n} d\omega \tag{5.8}$$

Considering the low-pass example given in Figure 5.4, the impulse response, obtained from Equation 5.8 is:

$$h_d(n) = \frac{\omega_c}{\pi} \frac{\sin(\omega_c n)}{\omega_c n}, -\infty \leq n \leq \infty \tag{5.9}$$

A rough sketch of the impulse response, given in Equation 5.9, is shown in Figure 5.5. On observing the impulse response in Figure 5.5, there are two fundamental problems:

- The impulse response $h_d(n)$ exists on both positive and negative sides of the time axis; hence, the system is *not causal.*
- The impulse response $h_d(n)$ exists to infinite extent on both sides of the time axis; hence, the system is *not finite.*

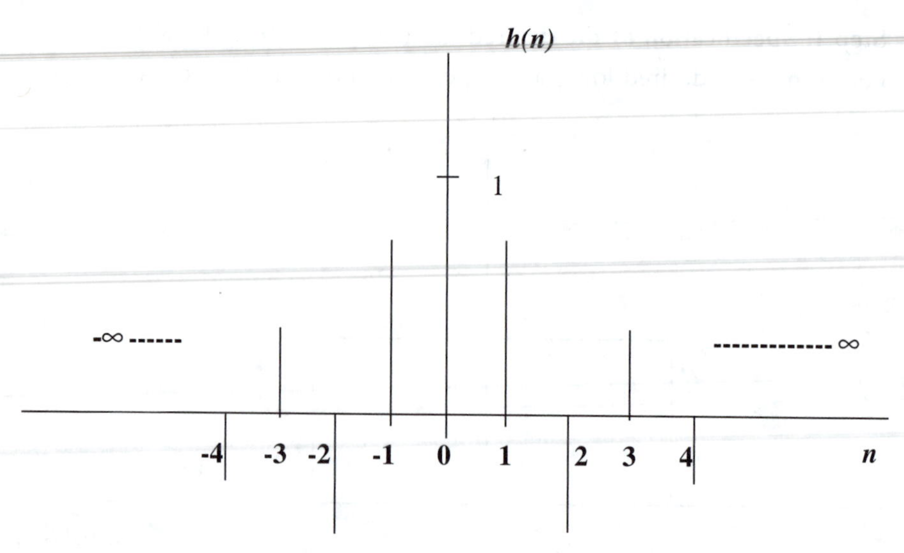


**FIGURE 5.5**
Impulse response of ideal low-pass filter.

**Step 3: Obtaining the causal and finite impulse response**

One of the simplest ways to obtain a realizable filter response $h(n)$, i.e., both finite and causal, from the ideal response $h_d(n)$, is to truncate the latter response by use of a window function $w(n)$ of length $M+1$, and to shift the filter impulse function by a value $\alpha = M/2$. This "windowing" operation is illustrated by the following equation:

$$h(n) = h_d(n - \alpha)w(n) \tag{5.10}$$

For example, *the rectangular window,* which is the simplest kind of window, is defined as:

$$w(n) = \begin{cases} 1, & 0 \le n \le M \\ 0, & \text{otherwise} \end{cases} \tag{5.11}$$

**Step 4: Frequency response of the windowed filter $H(e^{j\omega})$**

The frequency response of the windowed filter can be calculated either as:

$$H\left(e^{j\omega}\right) = \sum_{n=0}^{M} h(n)e^{-j\omega n} \tag{5.12}$$

or as the convolution integral:

$$H\left(e^{j\omega}\right) = \frac{1}{2\pi}\int_{-\pi}^{\pi} H_d\left(e^{j\theta}\right) W\left(e^{j(\omega-\theta)}\right) d\theta \tag{5.13}$$

**Step 5: Comparison of frequency response of the windowed filter $H(e^{j\omega})$ and the ideal desired filter $H_d(e^{j\omega})$**

The two frequency responses are compared in Figure 5.6. As is seen, the windowed filter frequency response is less ideal than the initial desired filter response. Some non-ideal properties of the filter include passband and stopband ripple in the frequency response and also a nonzero transition band. *However, the significant fact is that the windowed filter response can be optimized by appropriate choice of the window function.* This reason motivated several researchers in the field to design various types of window functions.

***Examples of commonly used windows***

Some of the commonly used windows include *Bartlett, Hanning, Hamming,* and *Kaiser windows,* whose equations are given below. The Kaiser window,

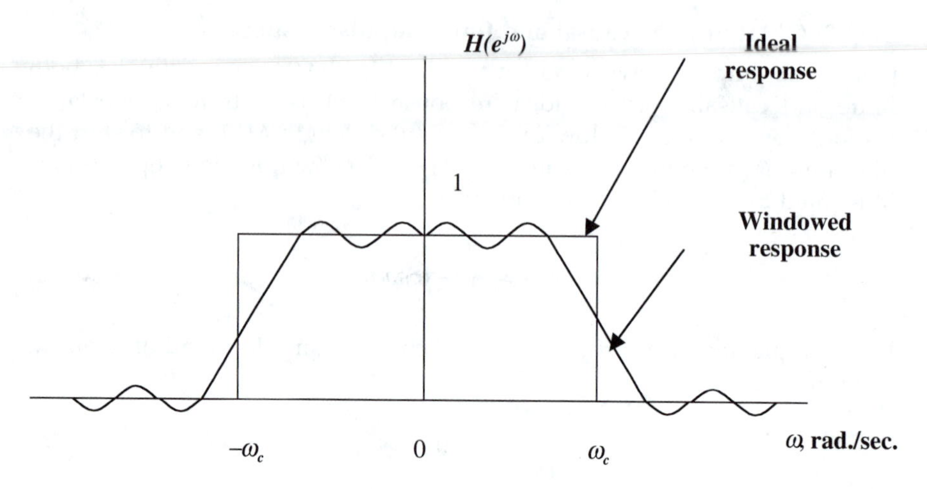


**FIGURE 5.6**
Ideal and windowed frequency response of low-pass filter.

however, also gives a practical design procedure to estimate the order of the filter. The order of the filter, *M*, is calculated based on desired specifications of the filter frequency response, which comprise the maximum tolerable ripple and the maximum tolerable transition bandwidth.

- Bartlett (triangular) window

$$w(n) = \begin{cases} 2n/M, & 0 \le n \le M/2 \\ 2 - 2n/M, & M/2 < n \le M \\ 0, & \text{otherwise} \end{cases} \quad (5.14a)$$

- Hanning window

$$w(n) = \begin{cases} 0.5 - 0.5\cos\left(2\pi n/M\right), & 0 \le n \le M \\ 0, & \text{otherwise} \end{cases} \quad (5.14b)$$

- Hamming window

$$w(n) = \begin{cases} 0.54 - 0.46\cos\left(2\pi n/M\right), & 0 \le n \le M \\ 0, & \text{otherwise} \end{cases} \quad (5.14c)$$

- Kaiser window

$$w(n)=\begin{cases}\dfrac{I_0\left[\beta\sqrt{\left\{1-\left(\dfrac{n-\alpha}{\alpha}\right)^2\right\}}\right]}{I_0(\beta)}, & 0\le n\le M\\ 0, & \text{otherwise}\end{cases} \tag{5.14d}$$

where $\alpha = M/2$ and $I_0$ is the modified Bessel function first kind and order zero.

The shape factor, $\beta$, can be adjusted to optimize the window properties, for certain desired properties of the filter frequency response. For a given maximum frequency response ripple, $\delta$, and maximum transition bandwidth $\Delta\omega$, the shape factor is given by the following equation:

$$\beta=\begin{cases}0.1101(A-8.7), & A>50\\ 0.5842(A-21)^{0.4}+0.07886(A-21), & 21\le A\le 50\\ 0, & A<21\end{cases} \tag{5.15}$$

where $A = -20\log_{10}\delta$.

Additionally, the order of the Kaiser window $M$ is given by the following empirical equation:

$$M=\frac{A-8}{2.285\Delta\omega} \tag{5.16}$$

Finally the impulse response of the windowed causal, finite filter is given by

$$h(n)=\begin{cases}h_d(n-\alpha)\dfrac{I_0\left[\beta\sqrt{\left\{1-\left(\dfrac{n-\alpha}{\alpha}\right)^2\right\}}\right]}{I_0(\beta)}, & 0\le n\le M\\ 0, & \text{otherwise}\end{cases} \tag{5.17}$$

*Note:* All window functions are symmetric about the point $M/2$. This implies the following condition:

$$w(n)=\begin{cases}w(M-n), & 0\le n\le M\\ 0, & \text{otherwise}\end{cases} \tag{5.18}$$

## 5.2 Problem Solving

**Exercise 1: Solve the following problems, briefly outlining the important steps**

a. The frequency response of a certain class of digital filters called binomial filters is written as

$$H_r(e^{j\omega}) = 2^N [\sin(\omega/2)]^r [\cos(\omega/2)]^{N-r}$$

in the range $-\pi \leq \omega \leq \pi$. Selecting $N = 2$, approximately sketch the response of the filters in the range $0 \leq \omega \leq \pi$ for the following cases:

i. $r = 0$

ii. $r = 1$

iii. $r = 2$

b. A 3-point symmetric moving average discrete-time filter is of the form:

$$y(n) = b[a\, x(n-1) + x(n) + a\, x(n+1)]$$

where $a$ and $b$ are constants. Determine, as a function of $a$ and $b$, the frequency response $H(e^{j\omega})$ of the system.

c. We wish to use the Kaiser window method to design a real-valued FIR filter that meets the following specifications:

$$0.9 < H\left(e^{j\omega}\right) < 1.1, \qquad 0 \leq |\omega| \leq 0.2\pi$$

$$-0.06 < H\left(e^{j\omega}\right) < 0.06, \quad 0.3\pi \leq |\omega| \leq 0.475\pi$$

$$1.9 < H\left(e^{j\omega}\right) < 2.1, \qquad 0.525\pi \leq |\omega| \leq \pi$$

The ideal frequency response $H_d(e^{j\omega})$ is given by

$$H_d(e^{j\omega}) = \begin{cases} 1, 0 \leq |\omega| \leq 0.25\pi \\ 0, 0.25\pi \leq |\omega| \leq 0.5\pi \\ 2, 0.5\pi \leq |\omega| \leq \pi \end{cases}$$

i. What is the maximum value of $\delta$ that can be used to meet this specification? What is the corresponding value of $\beta$?

ii. What is the maximum value of $\Delta\omega$ that can be used to meet this specification? What is the corresponding value of $M$?

d. We wish to use the Kaiser window method to design a digital band-pass filter satisfying the following specification:

$$-0.02 < H\left(e^{j\omega}\right) < 0.02, \quad 0 \le |\omega| \le 0.2\pi$$
$$0.95 < H\left(e^{j\omega}\right) < 1.05, \quad 0.3\pi \le |\omega| \le 0.7\pi$$
$$-0.001 < H\left(e^{j\omega}\right) < 0.001, \quad 0.75\pi \le |\omega| \le \pi$$

The filter will be designed by applying bilinear transform with $T$ = 5 ms to a prototype continuous-time filter. State the specifications that should be used to design a prototype continuous-time filter.

## 5.3 Computer Laboratory: Design of FIR and IIR Digital Filters Using Computer Aided Design (CAD) Techniques

This section explains how computer software, such as MATLAB,[4] can be used to design and implement FIR and IIR digital filters. Try out each of the commands given below and familiarize yourself with the types of MATLAB commands and formats.

### 5.3.1 Basic MATLAB Commands to Calculate and Visualize Complex Frequency Response

The difference equation of a general digital filter can be written as:

$$a_0 y(n) + a_1 y(n-1) + \ldots a_N y(n-N) = b_0 x(n) + b_1 x(n-1) + \ldots b_M x(n-M)$$

A compact MATLAB program is given below to plot the magnitude and phase responses of the filter system, defined above by the difference equation.

```
% MATLAB Program to Plot Magnitude and Phase Response
                    of a Digital Filter

p = 100;
a = [a0 a1 … aN];
b = [b0 b1 … bM];
[H,w] = freqz(b,a,p)   ;returns the p-point complex
                        frequency response.
subplot(2,1,1)
plot(w,abs(H))         ; plots the magnitude response
subplot(2,1,2)
plot(w,angle(H))       ; plots the phase response
```

- **H** is the complex frequency response vector
- **w** is a vector containing the $p$ frequency points in the range $0 \leq w \leq \pi$ radians.
- **a** and **b** are row vectors containing the coefficients $a_n$, $n = 0, 1, 2 \ldots N$ and $b_n$, $n = 0, 1, 2 \ldots M$

### 5.3.2 CAD of FIR Filters

#### *Method I. Window-based FIR filter design I*

```
>> b = fir1(N,wn,window); implements windowed low-pass
                          FIR filter design
```

- $b$ is a row vector containing the $N + 1$ coefficients of the order $N$ lowpass linear phase FIR filter with cutoff frequency $w_n$. The filter coefficients are ordered in descending shift order:

$$y(n) = b_0 x(n) + b_1 x(n-1) + \ldots b_M x(n-M)$$

- $w_n$ is the normalized cutoff frequency (normalized to $\pi$) and is a number between 0 and 1. If $w_n$, the cutoff frequency, is a 2-element vector $w_n = [w_1\ w_2]$, then *fir1* returns a *bandpass filter* with passband $w_1 < w < w_2$.
- $N$ is the order of the filter.
- *Window* is a column vector containing $N + 1$ elements of the specified window function $w(n)$. If no window is specified, *fir1* employs the Hamming Window.
- *High-pass filters* are designed by including the string *high* as a final argument.

  ```
  >> b = fir1(N,wn,'high', window)
  ```
- *Bandstop filters* are designed by including the string **stop** as the final argument and by specifying $w_n$ as a 2-element vector $w_n = [w_1\ w_2]$.

  ```
  >> b = fir1(N,wn,'stop', window)
  ```

#### *Method II. Window-based FIR filter design II*

```
>> b = fir2(N,f,H,window)
```

The *fir2* command designs digital filters with arbitrarily shaped response. This is in contrast to *fir1*, which designs filters in only standard low-pass, high-pass, bandpass, and bandstop configurations.

- ***b*** is a row vector containing the $N + 1$ coefficients of the order $N$ FIR filter, whose frequency magnitude characteristics are given by the vectors ***f*** and ***H***.
- ***f*** is a vector of frequency points, specified in the normalized range $0 \le f \le 1$, which corresponds to the digital frequency limit $0 \le \omega \le 1$.
- ***H*** is a vector containing the desired magnitude response at the points specified in the vector ***f***.

### *Method III. Optimization approach*

The most widely used program is the *Parks-McClellan Algorithm*.[5] The design procedure is as follows.

The difference equation of a *Nth order FIR* filter can be written as:

$$y(n) = b_0 x(n) + b_1 x(n-1) + \ldots b_M x(n-M)$$

Let $H_d(\omega)$ be the desired real-valued response of the FIR filter of order $m$. The error term $E(\omega)$ is defined as

$$E(\omega) = W(\omega)\left[H_d(\omega) - H(\omega)\right], \ -\pi \le \omega \le \pi$$

where $W(\omega)$ is a weighting factor.

If the designer attaches greater importance to the filter performance in a certain range of frequencies, then the weighting factor is higher in those frequency bands. The *optimization problem* is then stated as:

$$\underset{b(n)}{\text{Minimize}}\left[\text{Max}\left|E(\omega)\right|\right] \text{ in the range } -\pi \le \omega \le \pi$$

which means that the values of the filter coefficients $b(n)$, $n = 0, 1, \ldots N$ are to be chosen to minimize the maximum value of the error $E(\omega)$, $-\pi \le \omega \le \pi$. This optimization is done by the *Remez Exchange Method*. In MATLAB, the command

```
>> b = remez(N,f,H)
```

returns a linear phase filter with the $(N + 1)$ coefficients $b(n)$, $n = 0, 1, \ldots N$.

- The coefficients in the vector ***b*** are real and symmetric.
- ***f*** is a vector of frequency points, specified in the range $0 \le f \le 1$, which corresponds to the digital frequency limit $0 \le \omega \le 1$.
- The length of ***f*** and ***H*** must be the same and should be an even number.
- ***H*** is a vector containing the desired magnitude response at the points specified in the vector ***f***.

A MATLAB program which implements the *Remez Exchange Algorithm* is given below,

```
% MATLAB Program to Implement Remez Algorithm
N of order 20;

f = [0.1 0.2 0.3 0.4 0.5 0.6 0.7 0.8 0.9 1.0]

w = f*pi

Hd = [0 0 1 1 0 0 1 1 0 0 0]

b = remez(N,f,Hd);            yields the filter coefficients
                              b(n), n = 0,1,2 … M
[H w1] = freqz(b,1,128);      gives the actual filter
                              response H(w)
plot(w,Hd,w1,abs(H));         plots the desired and actual
                              filter response
```

### 5.3.3 CAD of IIR Filters

The difference equation of an *Nth order IIR* filter can be written as:

$$a_0y(n)+a_1y(n-1)+\ldots a_Ny(n-N)=b_0x(n)+b_1x(n-1)+\ldots b_Mx(n-M)$$

Let $H_d(\omega)$ be the desired real-valued response of the IIR filter of order $m$. The error term $E(\omega)$ is defined as:

$$E(\omega) = H_d(\omega) - H(\omega),\ -\pi \le \omega \le \pi$$

where $H(\omega)$ is the actual filter response.

The *optimization problem* is then stated as:

$$\underset{a(n),b(n)}{\text{Minimize}} \int_{-\pi}^{\pi} |E(\omega)|^2 \, d\omega$$

which means that the values of the filter coefficients $a(n)$, $n = 0, 1 \frac{1}{N} - N$ and $b(n)$, $n = 0, 1, \ldots N$ are to be chosen to minimize the mean squared error. This optimization is done by the *Yule-Walker Method.* In MATLAB, the command

```
>> [b,a] = yulewalk(N,f,H)
```

returns an Nth order IIR filter design with the $(N + 1)$ coefficients $a(n)$ and $b(n)$, n = 0, 1, … $N$.

- $f$ is a vector of frequency points, specified in the range $0 \le f \le 1$, which corresponds to the digital frequency limit $0 \le \omega \le 1$.
- $H$ is a vector containing the desired magnitude response at the points specified in the vector $f$.
- $N$ is the order of the filter.

**Exercise 2: Conversion of analog to digital filters**

There are two important methods of conversion of classical analog filter response $H(s)$ to corresponding digital filter response $H(z)$: *the impulse invariance method* and the *bilinear transformation.*

The analog transfer function is given by

$$H(s) = \frac{b_m s^m + b_{m-1} s^{m-1} + \ldots b_1 s + b_0}{a_n s^n + a_{n-1} s^{n-1} + \ldots a_1 s + a_0}$$

and the digital transfer function is given by

$$H(z) = \frac{b_m z^{-m} + b_{m-1} z^{-(m-1)} + \ldots b_1 z^{-1} + b_0}{a_n z^{-n} + a_{n-1} z^{-(n-1)} + \ldots a_1 z^{-1} + a_0}$$

Transform the following second-order cascade lowpass analog filter into digital filters (impulse invariance and bilinear methods):

$$H(s) = \frac{\Omega_c^2}{\left[s + \Omega_c\right]^2}$$

where $\Omega_c$ is the 3 dB cutoff frequency of the analog filter.

Design the digital filter cutoff frequency at $\omega_c = \pi/2$ rad./sec., and sampling frequency $f_s = 10$ Hz. Plot the magnitude response $|H(e^{j\omega})|$, $-\pi \le \omega \le \pi$ for both bilinear and impulse invariance transformation.

*Note:* The MATLAB commands are **bilinear** (for bilinear transformation) and **impinvar** (for impulse invariance). Type **help bilinear** or **help impinvar**, after the MATLAB prompt >> for instructions on usage.

**Exercise 3: Design of FIR filters using windowing method**

Design a digital windowed bandpass FIR filter of order 7 with the following desired frequency response:

$$H_d\left(e^{j\omega}\right) = 1, \quad \pi/3 \le |w| \le 2\pi/3$$

$$= 0, \text{ otherwise in the period } (-\pi, \pi)$$

Plot the desired and actual frequency responses for the following windowing functions:

a. Rectangular
b. Triangular
c. Hanning
d. Chebyshev (with sidelobe level of 30 dB below mainlobe)

In each case, record the following parameters in the actual frequency response:

i. Peak value of ripple (dB) in the passband
ii. Transition bandwidth in Hz.

**Exercise 4: Design of FIR and IIR filters using optimization techniques**

Design the bandpass filter of order 7 given in Exericse 3 using the following optimization methods:

a. FIR realization using the Remez Algorithm
b. IIR realization using the Yule-Walker Algorithm.

Plot the desired and actual frequency responses and in each case record the following parameters in the actual frequency response:

i. Peak value of ripple (dB) in the passband
ii. Transition bandwidth in Hz.

## References

1. Lathi, B.P., *Modern Digital and Analog Communication Systems,* 3rd Edition, Oxford University Press, New York, 1998.
2. Akay, M., *Biomedical Signal Processing,* Academic Press, San Diego, CA, 1994.
3. Nilsson, J.W. and Riedel, S.A., *Electric Circuits,* 6th Edition, Addison Wesley, Boston, MA, 2001.
4. *Student Edition of MATLAB/Simulink,* Mathworks, Natick, MA.
5. Mclellan, J.H. et al., A computer program for designing optimum FIR linear-phase digital filters, *IEEE Trans. on Audio and Electroacoustics,* Dec. 1983, 506–526.
6. Oppenheim, A.V and Schafer, R.W., with Buck, J.R., *Discrete-Time Signal Processing,* 2nd Edition, Prentice Hall, Upper Saddle River, NJ, 1998.
7. Kumar, B.P., *Digital Signal Processing Laboratory,* California State University, Sacramento, 2003.

# 6

# Digital Filter Design II: Applications

## 6.1 Introduction to Digital Filtering Applications

In the previous chapter, we covered the basic techniques of digital filtering, which included analytical and CAD methods for FIR/IIR digital filters. This chapter will focus on practical applications of digital filtering, involving the use of software tools that were discussed in the previous chapter. Typical applications of filters, which were explained briefly in the previous chapter, include frequency selection, signal demodulation, filtering of noisy audio and video signals, and time/frequency analysis of widely used signals such as the EKG (heart) and EEG (brain). In order to effectively carry out the video filtering exercise in this chapter, a brief overview of digital video processing will be presented.[1]

### 6.1.1 Brief Introduction to Digital Video Processing

A one-dimensional signal $x(t)$ is a function of one independent variable, $t$, or time, as in a speech signal. A two-dimensional signal $f(x,y)$ is a function of two independent variables, $x$ and $y$, which are usually the coordinates of space and are called *spatial variables*. Examples of two-dimensional spatial signals are images (photographic, infrared, or ultrasound), as shown in Figure 6.1. The function $f(x,y)$ represents the intensity of the image at the point $(x,y)$. For example, in the black-and-white image of Figure 6.1, the range of the function $f(x,y)$ would vary from 0 (black) to 1 (white) in a normalized intensity scale.

***Two-dimensional discrete signals***

Two-dimensional discrete signals are obtained by *sampling* two-dimensional continuous signals. A general point in the sampling grid is $(n_1\Delta x, n_2\Delta y)$, and the sampled signal is $f(n_1\Delta x, n_2\Delta y)$, or simply $f(n_1, n_2)$ in the range $0 \leq n_1 \leq N_1 - 1$; $0 \leq n_2 \leq N_2 - 1$. The sampled signal can be represented by the matrix function:

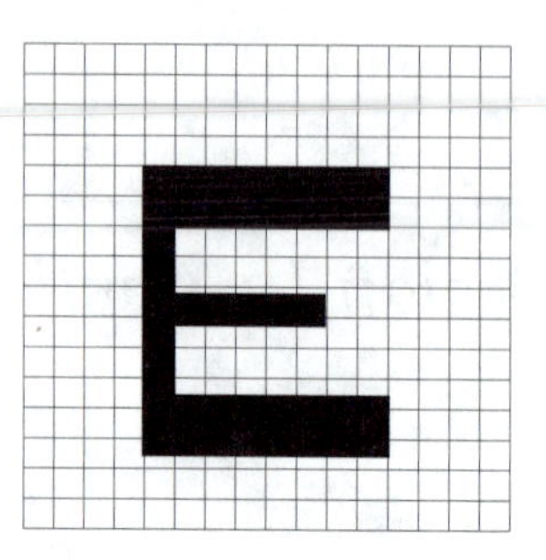

**FIGURE 6.1**
16 × 16 pixel image of the letter E.

$$f = \begin{bmatrix} f(0,0) & f(0,1)\ldots & f(0,N_2-1) \\ f(1,0) & f(1,1)\ldots & f(1,N_2-1) \\ \cdot & & \\ \cdot & & \\ \cdot & & \\ f(N_1-1,0) & f(N_1-1,1)\ldots & f(N_1-1,N_2-1) \end{bmatrix} \tag{6.1}$$

Each element of the matrix $f$ can also be termed a *pixel*, giving a total of $N_1$ at $N_2$ pixels in the entire image. Some common examples of two-dimensional discrete signals are:

- 2-d impulse function

$$\delta(n_1,n_2) = \begin{cases} 1, & \text{for } n_1 = n_2 = 0 \\ 0, & \text{otherwise} \end{cases} \tag{6.2}$$

- 2-d unit step function

$$u(n_1,n_2) = \begin{cases} 1, & \text{for } n_1 \geq 0, n_2 \geq 0 \\ 0, & \text{otherwise} \end{cases} \tag{6.3}$$

***Two-dimensional discrete systems***

A system with two-dimensional discrete space input and output signals is termed a 2-*d discrete-space system,* as shown in Figure 6.2. The relationship between the output and input of a 2-d discrete system is given by

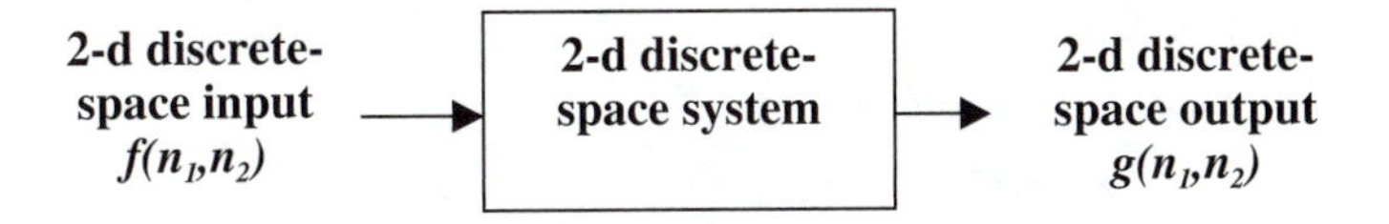


**FIGURE 6.2**
Two-dimensional discrete space system.

$$g(n_1,n_2)=T\left[f(n_1,n_2)\right] \tag{6.4}$$

where $T$ is the system operator.

If the system is *LSI* (*Linear Shift Invariant*), then we have the 2-d convolution relation:

$$g(n_1,n_2)=\sum_{k_1}\sum_{k_2} f(k_1,k_2)h(n_1-k_1,n_2-k_2) \tag{6.5}$$

or

$$g(n_1,n_2)=f(n_1,n_2)**h(n_1,n_2) \tag{6.6}$$

where the symbol ** represents the 2-d discrete convolution, and $h(n_1, n_2)$ is the 2-d impulse response of the system.

The 2-d impulse response is defined as the output of the system, when the input $f(n_1, n_2) = \delta(n_1, n_2)$, the 2-d impulse function, defined in Equation 6.2.

***Two-dimensional Discrete-Time Fourier Transform (2-d DTFT)***

The 2-d DTFT of a 2-d discrete function $f(n_1, n_2)$ is defined as:

$$F(\omega_1,\omega_2)=\sum_{n_1}\sum_{n_2} f(n_1,n_2)e^{-j\omega_1 n_1}e^{-j\omega_2 n_2} \tag{6.7}$$

Likewise, the 2-d inverse DTFT is given by the following equation:

$$f(n_1,n_2)=\frac{1}{4\pi^2}\int_{-\pi}^{\pi}\int_{-\pi}^{\pi} F(\omega_1,\omega_2)e^{j\omega_1 n_1}e^{j\omega_2 n_2}d\omega_1\, d\omega_2 \tag{6.8}$$

*Note:* The 2-d Fourier Transform $F(\omega_1,\omega_2)$ is periodic in both variables $\omega_1$ and $\omega_2$ with period $2\pi$ radians. This implies that

$$F(\omega_1 \pm k_1 2\pi,\omega_2 \pm k_2 2\pi)=F(\omega_1,\omega_2);\ \text{for integer } k_1 \text{ and } k_2. \tag{6.9}$$

**TABLE 6.1**

2-d DTFT Theorems

| Property | $f(n_1, n_2)$ | $F(\omega_1, \omega_2)$ |
|---|---|---|
| Convolution | $f(n_1, n_2) ** h(n_1, n_2)$ | $F(\omega_1, \omega_2)\, H(\omega_1, \omega_2)$ |
| Spatial shift | $f(n_1 - m_1, n_2 - m_2)$ | $F(\omega_1, \omega_2)e^{-j\omega_1 m_1}\, e^{-j\omega_2 m_2}$ |
| Spatial reflection | $f(-n_1, n_2)$ | $F(-\omega_1, \omega_2)$ |

Some of the other important 2-d DTFT properties are given in Table 6.1.

### *2-d Discrete Fourier Transform (2-d DFT)*

In Section 3.1.2, we discussed the one-dimensional Discrete Fourier Transform (DFT) as the practical extension of the 1-d DTFT. Similarly, we can extend the DFT concept to the 2-d DTFT, $F(\omega_1, \omega_2)$, of the 2-d spatial signal $f(n_1, n_2)$. Because the 2-d DTFT $F(\omega_1, \omega_2)$ is periodic in both variables $\omega_1$ and $\omega_2$ with period $2\pi$ radians, this property is used to the divide the frequency interval $(0, 2\pi)$ into $N_1$ (for $\omega_1$) and $N_2$ (for $\omega_2$) equally spaced points. This discretization yields the 2-d Discrete Fourier Transform (2-d DFT) of the 2-d spatial signal $f(n_1, n_2)$ as follows:

$$F(k_1, k_2) = F(\omega_1, \omega_2)\Big|_{\substack{\omega_1 = 2\pi k_1/N_1 \\ \omega_2 = 2\pi k_2/N_2}} = \sum_{n_1=0}^{N_1-1}\sum_{n_2=0}^{N_2-1} f(n_1, n_2)e^{-j2\pi n_1 k_1/N_1}e^{-j2\pi n_2 k_2/N_2} \quad (6.10)$$

$$\text{for } k_1 = 0,\ 1,\ 2 \ldots N_1 - 1;$$

$$k_2 = 0,\ 1,\ 2 \ldots N_2 - 1;$$

Similarly the inverse 2-d DFT is defined as follows:

$$f(n_1, n_2) = \frac{1}{N_1 N_2}\sum_{k_1=0}^{N_1-1}\sum_{k_2=0}^{N_2-1} F(k_1, k_2)e^{j2\pi n_1 k_1/N_1}e^{j2\pi n_2 k_2/N_2} \quad (6.11)$$

$$\text{for } n_1 = 0,\ 1,\ 2 \ldots N_1 - 1;$$

$$n_2 = 0,\ 1,\ 2 \ldots N_2 - 1;$$

Both $f(n_1, n_2)$ and $F(k_1, k_2)$ are matrices of size $N_1 \times N_2$. Additionally, as seen from Equation 6.10 and Equation 6.11, the 2-d DFT and 2-d IDFT are both finite sums, making it very convenient to program in computers and microprocessors.

### *2-d Fast Fourier Transform*

The 2-d Fast Fourier Transform (2-d FFT) is an extension of the 1-d FFT discussed in Section 3.1.3. As in the 1-d case, it is a much faster computation method of the 2-d DFT discussed in the previous section, by using some periodic properties of the exponential functions in Equation 6.10 and Equation 6.11.

An $N_1 \times N_2$ point 2-d DFT or $N_1 \times N_2$ point 2-d IDFT requires $N_1^2 \times N_2^2$ complex multiplications, whereas a 2-d FFT requires only $(N_1 \log_2 N_1) \times (N_2 \log_2 N_2)$ complex multiplications, when both $N_1$ and $N_2$ are powers of 2, also termed as *radix-2 2-d FFT*. This is especially significant for large values of $N_1$ and $N_2$. For example, when $N_1 = N_2 = 128$, the number of complex multiplications are 268,435,460 for direct 2-d DFT computation, and only 229,376 for a radix-2 2-d FFT.

## 6.1.2 Simulation of 2-Dimensional Imaging Process

The fundamental aspect of 2-dimensional image processing is the transmission of discrete video signals through the medium, between the *transmitter* and *receiver*. Assuming the medium to be LSI, the complete imaging process, as shown in Figure 6.3, is described in the *spatial domain* by the following equations, in the spatial range $0 \le n_1 \le N_1 - 1$, $0 \le n_2 \le N_2 - 1$:

Restored image $$y(n_1, n_2) = x(n_1, n_2) ** h(n_1, n_2) + \eta(n_1, n_2) \tag{6.12}$$

and

Restored image $$x_r(n_1, n_2) = y(n_1, n_2) ** g(n_1, n_2) \tag{6.13}$$

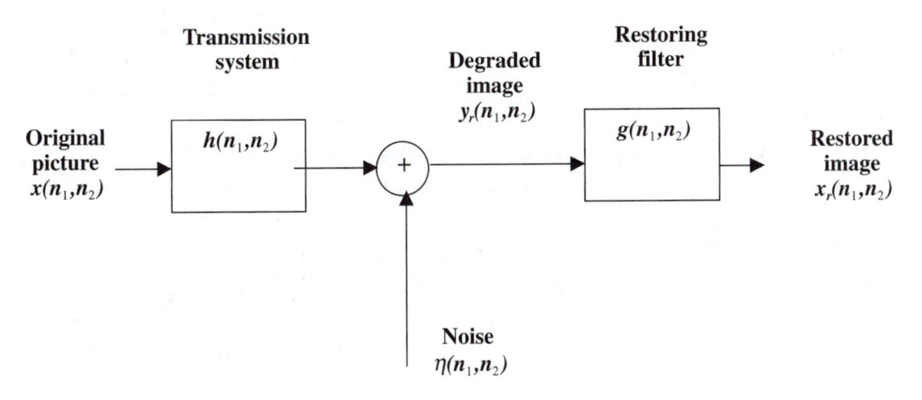


**FIGURE 6.3**
Basic imaging system.

where

$x(n_1, n_2)$ is the original 2-d digital picture,
$h(n_1, n_2)$ is the impulse response of the *transmission system*,
$\eta(n_1, n_2)$ is the additive *noise*,
$y(n_1, n_2)$ is the degraded image,
$g(n_1, n_2)$ is the impulse response of the *restoring filter*, and
$x_r(n_1, n_2)$ is the restored image.

In the *spatial frequency domain*, Equation 6.12 and Equation 6.13 become:

$$Y(\omega_1, \omega_2) = X(\omega_1, \omega_2)\, H(\omega_1, \omega_2) + N(\omega_1, \omega_2) \tag{6.14}$$

and

$$X_r(\omega_1, \omega_2) = Y(\omega_1, \omega_2)\, G(\omega_1, \omega_2) \tag{6.15}$$

The MATLAB program for implementing Equation 6.12 and Equation 6.13 is given below.

```
% MATLAB Program for Implementation of the 2-D Imaging
                and Restoration Process
>> for n1 = 1:1:N1
   for n2 = 1:1:N2
     x(n1, n2) =    ; defines the original digital picture
     h(n1, n2) =    ; defines the impulse response of the
                      system
     n(n1, n2) =    ; defines the additive noise
   end;
   end;
>> X = fft2(x)      ; calculates the 2-d DFT of x(n1, n2)
>> H = fft2(h)      ; calculates the 2-d DFT of h(n1, n2)
>> N = fft2(n)      ; calculates the 2-d DFT of n(n1, n2)
>> Y = X.*H + N     ; calculates Equation.6.14
>> G = fft2(g)      ; calculates the 2-d DFT of the
                      restoring filter impulse response
                      g(n1, n2)
>> Xr = Y.*G        ; calculates Equation 6.15
>> xr = ifft2(Xh) ;obtains the restored image matrix
                      xr(n1, n2)
>> xr               ;displays the restored image matrix
                      xr(n1, n2)
>> e = xr - x       ; obtains the error matrix e(n1, n2)
>> et = (norm(e)/norm(x))*100.; obtains the relative
                    error of transmission
```

## 6.2 Problem Solving

**Exercise 1: Solve the following problems, briefly outlining the important steps.**

a. Sample the following 2-d continuous functions $f(x,y)$ (in the interval given) to obtain 2-d discrete functions $f(n_1,n_2)$ in the form of $4 \times 4$ matrices. Sketch the sampled 2-d functions.

   i. $f(x,y) = rect(x/2,y/2)$, in the interval $-1 \leq \eta \leq 1$, $-1 \leq y \leq 1$,

$$\text{where } rect\left(x/a, y/b\right) = \begin{cases} 1, & -a \leq x \leq a;\ -b \leq x \leq b \\ 0, & \text{otherwise} \end{cases}$$

   ii. $f(x,y) = sin(\pi x/4)\, sin(\pi y/4)$, in the interval $0 \leq x \leq 8$, $0 \leq y \leq 8$.

b. Find the 2-d DTFT $F(\omega_1,\omega_2)$ of the following 2-d discrete functions $f(n_1,n_2)$:

   i. $f(n_1,n_2) = \delta(n_1 - 2,n_2 - 2) + \delta(n_1 - 1,n_2 - 3) + \delta(n_1 - 3,n_2 - 1)$

   ii. $f(n_1,n_2) = \delta(n_1 - 1)\, u(n_1,n_2)$

   iii. $f(n_1,n_2) = e^{-(n_1+n_2)}\, u(n_1,n_2)$

c. The block diagram of an LSI system is given in Figure 6.4:

$$\text{where } x(n_1, n_2) = (0.5)^{n_1} (0.25)^{n_2} u(n_1, n_2)$$

$$h(n_1, n_2) = \delta(n_1, n_2) + \delta(n_1 - 1, n_2) + \delta(n_1, n_2 - 1) + \delta(n_1 - 1, n_2 - 1)$$

   i. Determine the 2-d Fourier transform $H(\omega_1,\omega_2)$ of the system impulse response $h(n_1, n_2)$.

   ii. Determine the 2-d Fourier transform $Y(\omega_1,\omega_2)$, and hence, determine the output $y(n_1, n_2)$.

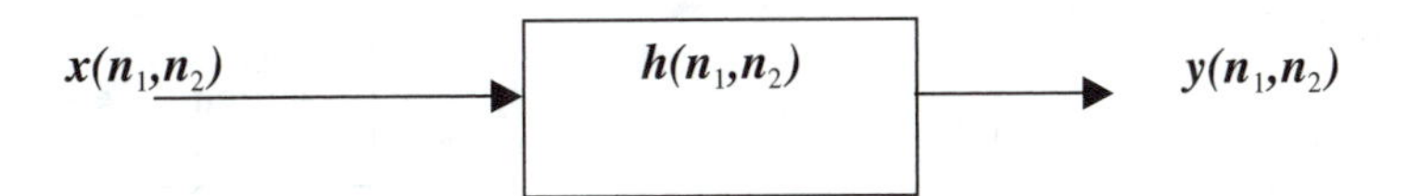


**FIGURE 6.4**
Figure for problem (c).

## 6.3 Computer Laboratory

This section consists of a set of exercises, based on practical applications of digital filtering. These exercises require predominant use of MATLAB or Simulink.

### 6.3.1 Frequency Selection Applications

**Exercise 2: Filtering of mixed sinusoidal signals of different frequency**

a. Create a new model file in either MATLAB or Simulink,[2] as shown in Figure 6.5. Generate a mixed signal consisting of two sinusoidal signals of frequency 2 KHz and 4 KHz. Verify the mixed output signal on both the oscilloscope and the FFT analyzer, if using Simulink. If using MATLAB, use the *fft* command to generate the output spectrum, according to the procedure for periodic signals detailed in Section 3.1.3.
b. Sample the combined signal, at the appropriate sample frequency, by passing it through a sample and hold circuit, and then pass the sampled signal through a *digital bandpass filter* centered at 2 KHz, and a bandwidth of 0.4 KHz, using the three-step procedure outlined in Section 5.1.2.2. Implement the filter using at least two of the methods described in Section 5.3.2, such as **fir1**, **fir2**, and **Remez** commands. Verify that the filtered output is predominantly the 2 KHz signal and note down the voltage levels of both the signal components.
c. Repeat the procedure described in the previous section, and implement a digital bandpass filter centered at 4 KHz, and a bandwidth of 0.4 KHz. As in the previous step, verify that the filtered output is

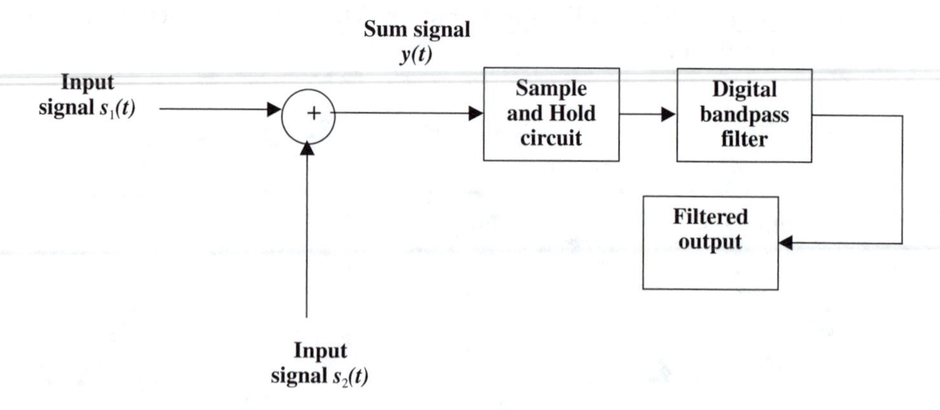


**FIGURE 6.5**
Model for filtering a combination of two sinusoidal signals.

predominantly the 4 KHz signal and note down the voltage levels of both the signal components.

### 6.3.2 Signal Demodulation Applications

**Exercise 3: Filtering of amplitude modulated (AM) signal to recover the baseband signal**

a. Simulate the AM *modulation* system,[3] as shown in Figure 6.6a, using MATLAB or Simulink. Check the AM output at the carrier, and two sideband frequencies.
b. Simulate the AM *demodulation* system, as shown in Figure 6.6b, using MATLAB or Simulink. The output of the multiplier will consist of both high frequency and low frequency signal components. Sample the multiplier output by passing it through the sample and hold circuit, at the Nyquist sampling rate corresponding to the highest frequency, which would be the higher sideband frequency.
c. Design a digital low-pass filter having a cutoff frequency slightly higher than 2 KHz, which is the frequency of the input baseband signal.

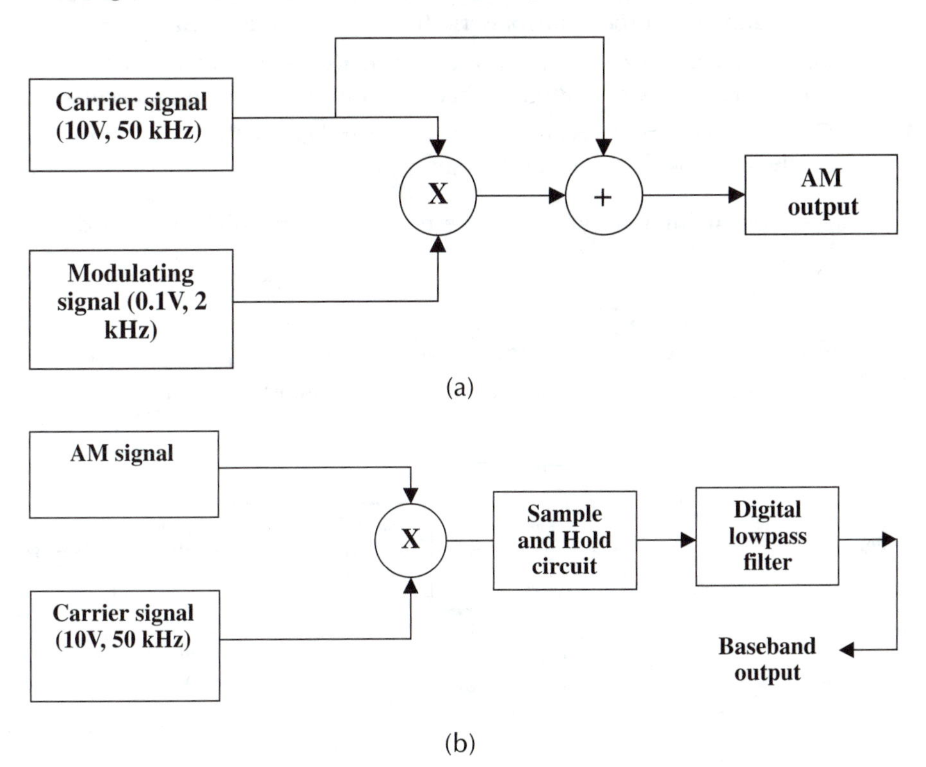


**FIGURE 6.6**
(a) Model for amplitude modulation system; (b) model for amplitude demodulation system.

Check the demodulated output on the scope and plot the error signal between the recovered baseband output and the input modulating signal. Determine the percentage mean-squared error (mse) between the two signals.

### 6.3.3 Filtering of Noisy Audio Signals

**Exercise 4: Filtering of one-dimensional time signals mixed with random noise**

Communication systems face the common problem of noise.[3] As shown in Figure 6.7, the simplest form of noise is additive noise $n(t)$, which adds on to the transmitted signal $s(t)$. Several methods have been developed to tackle the problem of noise removal from the corrupted signal $y(t) = s(t) + n(t)$. The commonly used methods include autocorrelation and filtering.

a. Create a new model file either in MATLAB or Simulink, as shown in Figure 6.7. Generate an analog sinusoidal signal at a frequency of 3 KHz and amplitude of 5 volts. Verify the sinusoidal output, $s(t)$, on both the oscilloscope and the FFT analyzer, if using Simulink. If using MATLAB, use the **fft** command to generate the output spectrum, according to the procedure for periodic signals detailed in Section 3.1.3.
b. Generate a uniform random noise signal, $n(t)$, with a signal-to-noise voltage ratio (SNR) of 30 dB. Check the output on the oscilloscope.
c. Combine the signal $s(t)$ and the noise $n(t)$ and check the noisy output on the oscilloscope and the FFT analyzer.

In this experiment, two types of noise-removal filters will be designed and tested.

- **Digital bandpass filter**

  Design a digital bandpass filter with a center frequency of 3 KHz, and suitable bandwidth to filter out the sinusoidal signal $s(t)$ from

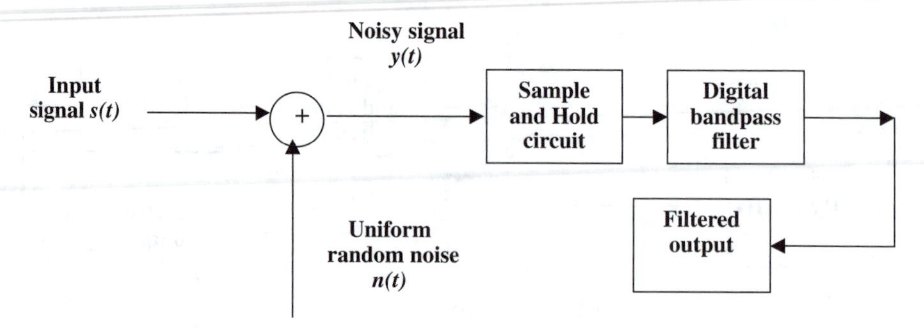


**FIGURE 6.7**
Model for filtering of noisy sinusoidal signal.

the noisy signal $y(t)$. Verify that the output of the filter is predominantly the 3 KHz signal.

- **Digital low-pass filter**

  Design a digital low-pass filter with cutoff frequency of 4 KHz, to filter out the sinusoidal signal $s(t)$ from the noisy signal $y(t)$. As in the bandpass case, verify that the output of the filter is predominantly the 3 KHz signal.

Repeat the simulation, using both digital bandpass filtering and digital low-pass filtering, for different noise levels, with SNR (voltage) of 20 dB and 10 dB, respectively.

### 6.3.4 Filtering of Noisy Video Signals

**Exercise 5: Filtering of two-dimensional spatial signals mixed with random noise**

A 2-d digital picture[1] representing the letter $E$ is transmitted through the system shown in Figure 6.3.

**Stage 1. Image Degradation**

a. Obtain a 16 × 16 picture matrix $x(n1, n2)$ consisting of 256 pixels, representing the letter $E$, with each pixel quantized to only 2 levels, 0 or 1. Please refer to Figure 6.1, which shows how to represent the letter $E$ in pixel format.

b. Obtain the 16 × 16 transmission matrix $h(n_1, n2)$ by sampling the following continuous function:

$$h(x,y) = \frac{e^{-jk\sqrt{x^2+y^2+z^2}}}{\sqrt{x^2+y^2+z^2}} \tag{6.16}$$

where the propagation constant $k$ = 1 m$^{-1}$ and the propagation distance $z$ = 5 m. Sample the function in the interval –8 m ≤ $x$ ≤ 7 m, –8 m ≤ $y$ ≤ 7 m (i.e., $\Delta x$ = 1m and $\Delta y$ =1m, if we have a 16 × 16 matrix). All spatial variables are defined in meters (m).

c. Obtain the degraded image matrix $y(n_1, n_2) = x(n_1, n_2) ** h(n_1, n_2) + \eta(n_1, n_2)$, where $\eta(n_1, n_2)$ is a 16 × 16 *random noise matrix* having a maximum value of 0.2. The random noise matrix is generated by the MATLAB command

```
>> M *rand(N); M is the maximum value of the noise and
               N is the order of the noise matrix
               (N = 16, in our case)
```

**Stage 2. Image Restoration**

a. The restoration is done by passing the degraded image $y(n1, n2)$ through a restoring filter $g(n1, n2)$. Determine the $16 \times 16$ restored image matrices $x1(n1, n2)$ and $x2(n1, n2)$, respectively, for the following filters:

- Inverse filter: $G(\omega_1,\omega_2)=\dfrac{1}{H(\omega_1,\omega_2)}$

- Wiener filter: $G(\omega_1,\omega_2)=\dfrac{H^*(\omega_1,\omega_2)}{\left|H(\omega_1,\omega_2)\right|^2+\dfrac{\left|N(\omega_1,\omega_2)\right|^2}{\left|X(\omega_1,\omega_2)\right|^2}}$

  where the symbol * denotes complex conjugate.

b. For comparison, determine the relative error of transmission $et = 100 \times \text{norm}(e)/\text{norm}(x)$ in the following cases:
   - Without any restoring filter: $e(n_1, n_2) = y(n_1, n_2) - x(n_1, n_2)$
   - With inverse filter: $e(n_1, n_2) = x_1(n_1, n_2) - x(n_1, n_2)$
   - With Wiener filter: $e(n_1, n_2) = x_2(n_1, n_2) - x(n_1, n_2)$

**Stage 3. Thresholding of Images and Display**

We can apply the thresholding process to an image $x(n1, n2)$ and obtain a display image $xt(n_1, n_2)$ as follows:

```
% MATLAB Program to Apply Thresholding on Filtered Image
 for n1= 1:16
 for n1= 1:16
   if abs(x(n1, n2)) > T xt(n1, n2) = '*'
   else xt(n1, n2) = ' '
   end
 end
 end
```

Apply the thresholding process to the following images, and display them by using a threshold level of $T = 0.5$.

- The original picture $x(n_1, n_2)$
- The restored image $x_1(n_1, n_2)$ obtained by inverse filtering
- The restored image $x_2(n_1, n_2)$ obtained by Wiener filtering

### 6.3.5 Image Compression Techniques

**Exercise 6: Image compression using transform coding; the Discrete Cosine Transform**

The Discrete Cosine Transform[1] (DCT) is the industry standard in image processing. This project will demonstrate the effectiveness of the DCT in image compression. The 2-d of an $N \times N$ pixel image $g(m,n)$, $m = 0, 1 \ldots N-1$, $n = 0, 1 \ldots N-1$ is given by

$$t(i,j) = c(i,j)\sum_{n=0}^{N-1}\sum_{m=0}^{N-1} g(m,n)\cos\frac{(2m+1)i\pi}{2N}\cos\frac{(2n+1)j\pi}{2N}$$

and the inverse DCT is calculated as follows:

$$g(m,n) = \sum_{i=0}^{N-1}\sum_{j=0}^{N-1} c(i,j)t(i,j)\cos\frac{(2m+1)i\pi}{2N}\cos\frac{(2n+1)j\pi}{2N}$$

where the coefficients $c(i,j)$ are given as: $c(0,j) = 1/N$, $c(i,0) = 1/N$ and $c(i,j) = 2/N$, for $i, j \geq 0$.

a. Write a MATLAB program to implement an $N \times N$-point DCT and inverse DCT.

b. Select an $N \times N$ square image (maximum value of $N$ is 256) from the MATLAB library and save it into the image matrix $g(m,n)$ as follows:

```
>> g = imread('filename.jpg')
```

The matrix $g$ will be of size $N \times N$ and contain real valued numbers which represent the gray level in each pixel. For color pictures the image is stored in a 3-dimensional matrix of size $N \times N \times 3$, with one $N \times N$ submatrix for each of the primary colors — red, blue, and green. The image can again be displayed on the screen by the command:

```
>> image(g)
```

***Image Compression Technique***

c. Take the 2-d DCT of the image $g(m,n)$ to obtain the transformed $N \times N$ matrix $t(i,j)$. The significance of the DCT matrix $t(i,j)$ is that it can be compressed (using different techniques) by setting elements in the matrix $t(i,j)$ to zero. This creates a new transform $N \times N$ matrix $t'(i,j)$, which has one or more zero elements. The inverse DCT of the matrix $t'(i,j)$ will yield the reconstructed image matrix $g'(m,n)$. The mean squared reconstruction error ($E_{mse}$) is defined as:

$$E_{mse} = \frac{1}{N^2} \sum_{n=0}^{N-1} \sum_{m=0}^{N-1} \left| g(m,n) - g'(m,n) \right|^2$$

In this project, two compression techniques will be attempted:

- ***Low pass filtering***: One common property of all DCT transform matrices is that the major part of the energy of $t(i,j)$ is concentrated in one corner ($0 \le m, n < N_1$) of the matrix.

d. Create new transformed matrices $t'(i,j)$, choosing $N_1 = N/2$ and $N/4$, putting the remaining values of the matrix $t(i,j)$ to zero.

e. Plot the reconstructed image and determine the $E_{mse}$ in each case.

- ***Selective matrix truncation***: In this compression method, the original transform matrix $t(i,j)$ is truncated by setting individual matrix elements to zero, one at a time. The pixels with the least variance are removed first, and then the pixels with increasing variance. The variance of the element $t(i,j)$ is defined as follows:

$$\text{Var}\left[ t(i,j) \right] = \sum_{k=1}^{<=8} \left| t(i,j) - t_k(i,j) \right|^2$$

where $t_k(i,j)$ is one of the eight neighboring pixels. For elements on the borders of the matrix, the neighboring elements will be less than eight.

f. Implement this truncation method, one pixel at a time, and determine the $E_{mse}$ between the reconstructed image and the original image in each case. Make a plot of the $E_{mse}$ versus the number of pixels removed. At which stage of the process is the image just visible? Plot the image at that stage only.

### 6.3.6 Time-Frequency Analysis of Practical Signals

**Exercise 7: Spectral analysis of the electrocardiogram (EKG) signal**

The record of the potential fluctuations during the cardiac cycle of the heart is called the electrocardiogram (EKG or ECG).[4] Most EKG machines record these fluctuations on a moving strip of paper. The EKG is a very useful means of diagnosing abnormalities in the heart by analyzing the EKG waveform directly on a time axis, or by analyzing the Fourier spectrum of the EKG recording on a frequency axis.

The EKG is a periodic time signal, as shown in Figure 6.8, for a normal heart. It has a period of ~1 second, corresponding to a fundamental frequency of ~1 Hz. It is characterized by the PQRSTU peaks as shown in Figure 6.8. In this exercise, we consider the following two types of heart conditions.

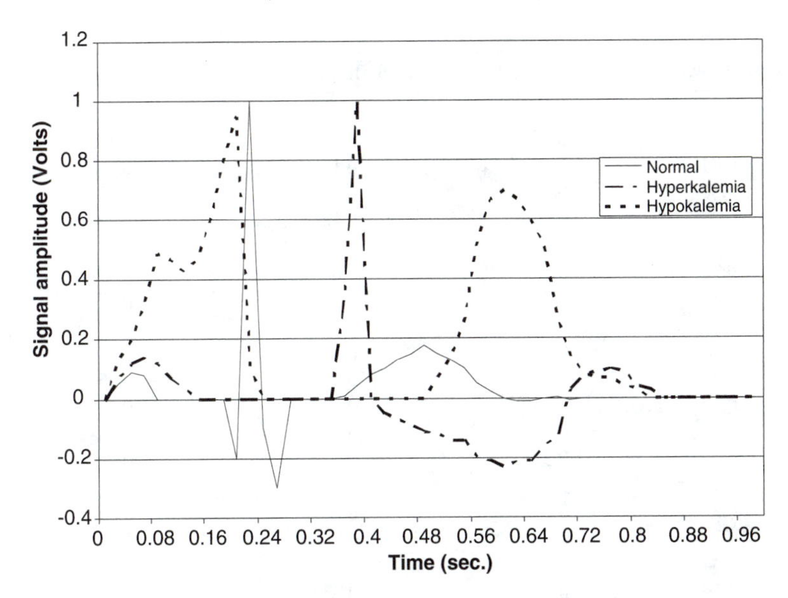


**FIGURE 6.8**
EKG patterns for normal, hypokalemia, and hyperkalemia cases.

**Hypokalemia**: The normal potassium level in the human body is in the range of 4 to 5.5 meq./liter. However, when the potassium level is lower than normal, the condition is called hypokalemia, and the EKG recording of a patient with this condition is also shown in Figure 6.8.

**Hyperkalemia**: When the potassium level is higher than normal, the condition is called hyperkalemia, and the EKG recording of a patient with this condition is shown in Figure 6.8.

a. Sample each of the signals, shown in Figure 6.8, over a time period of 1 sec. The numerical values of the EKG patterns, shown in Figure 6.8, are given in Table 6.2. Use a value of $N$ (number of sampling points) as 16 or higher to obtain good resolution. Use zero-padding if necessary, in order to use a radix-2 FFT.

b. Determine the FFT spectrum of each of the sampled signals, using the procedure detailed in Section 3.1.3. Plot the FFT magnitude and phase spectra of each of the signals.

c. Compare the magnitude spectrum of the FFTs obtained in the previous step and comment on the differences in the spectrum that will enable the user to distinguish between normal, hypokalemia, and hyperkalemia conditions.

**TABLE 6.2**

EKG Data

| Time (Sec.) | Normal (Volts) | Hyperkalemia(Volts) | Hypokalemia(Volts) |
|---|---|---|---|
| 0 | 0 | 0 | 0 |
| 0.02 | 0.05 | 0.07 | 0.135 |
| 0.04 | 0.09 | 0.12 | 0.2 |
| 0.06 | 0.08 | 0.14 | 0.32 |
| 0.08 | 0 | 0.12 | 0.49 |
| 0.1 | 0 | 0.07 | 0.47 |
| 0.12 | 0 | 0.04 | 0.43 |
| 0.14 | 0 | 0 | 0.47 |
| 0.16 | 0 | 0 | 0.6 |
| 0.18 | 0 | 0 | 0.8 |
| 0.2 | –0.2 | 0 | 0.95 |
| 0.22 | 1 | 0 | 0.1 |
| 0.24 | –0.1 | 0 | 0 |
| 0.26 | –0.3 | 0 | 0 |
| 0.28 | 0 | 0 | 0 |
| 0.3 | 0 | 0 | 0 |
| 0.32 | 0 | 0 | 0 |
| 0.34 | 0 | 0 | 0 |
| 0.36 | 0.01 | 0.33 | 0 |
| 0.38 | 0.04 | 1 | 0 |
| 0.4 | 0.08 | 0 | 0 |
| 0.42 | 0.1 | –0.05 | 0 |
| 0.44 | 0.13 | –0.07 | 0 |
| 0.46 | 0.15 | –0.09 | 0 |
| 0.48 | 0.18 | –0.11 | 0 |
| 0.5 | 0.15 | –0.12 | 0.07 |
| 0.52 | 0.13 | –0.14 | 0.14 |
| 0.54 | 0.1 | –0.14 | 0.26 |
| 0.56 | 0.05 | –0.2 | 0.53 |
| 0.58 | 0.02 | –0.21 | 0.67 |
| 0.6 | 0 | –0.23 | 0.7 |
| 0.62 | –0.01 | –0.21 | 0.67 |
| 0.64 | –0.01 | –0.21 | 0.6 |
| 0.66 | 0 | –0.16 | 0.52 |
| 0.68 | 0.005 | –0.13 | 0.27 |
| 0.7 | –0.005 | 0.03 | 0.14 |
| 0.72 | 0 | 0.08 | 0.08 |
| 0.74 | 0 | 0.09 | 0.07 |
| 0.76 | 0 | 0.1 | 0.07 |
| 0.78 | 0 | 0.09 | 0.04 |
| 0.8 | 0 | 0.04 | 0.03 |
| 0.82 | 0 | 0.03 | 0 |
| 0.84 | 0 | 0 | 0 |
| 0.86 | 0 | 0 | 0 |
| 0.88 | 0 | 0 | 0 |
| 0.9 | 0 | 0 | 0 |
| 0.92 | 0 | 0 | 0 |
| 0.94 | 0 | 0 | 0 |
| 0.96 | 0 | 0 | 0 |
| 0.98 | 0 | 0 | 0 |

## References

1. Haddad, R.A. and Parsons, T.W., *Digital Signal Processing — Theory, Applications, and Hardware*, Computer Science Press, Rockville, MD, 1991.
2. *Student Edition of MATLAB/Simulink*, Mathworks, Natick, MA. Version 5.3, 1999.
3. Lathi, B.P., *Modern Digital and Analog Communication Systems*, Third Edition, Oxford University Press, New York, 1998.
4. Akay, M., *Biomedical Signal Processing*, Academic Press, San Diego, CA, 1994.
5. Oppenheim, A.V and Schafer, R.W., with Buck, J.R., *Discrete-Time Signal Processing*, 2nd Edition, Prentice Hall, Upper Saddle River, NJ, 1998.
6. Kumar, B.P., *Digital Signal Processing Laboratory*, California State University, Sacramento, 2003.

# 7

# *DSP Hardware Design I*

## 7.1 Background of Digital Signal Processors

Digital signal processing (DSP) is a rapidly growing field within electrical and computer engineering. Analog processing is achieved using components such as resistors, capacitors, and inductors, whereas digital processing uses a programmable microprocessor. The main advantage of digital processing is that applications can be changed, corrected, or updated very easily by reprogramming the microprocessor, unlike analog systems, which would require components, such as resistors or capacitors, to be physically changed. Additionally, DSPs also reduce noise, power consumption, and cost, when compared with analog systems.

With processing power doubling every 18 months (according to Moore's law), the number of applications suitable for DSP is increasing at a comparable rate. In this introductory lab on DSP hardware, we will be using the Texas Instruments (TI) digital signal processors (DSPs), and the aim of the lab is to become familiar with the essential tools to set up and program the processors for practical applications. We will briefly discuss some important details of the DSPs before proceeding with the actual experiments. More detailed information on TI DSPs is available in Appendix E of this book and in the references.[1–10]

### 7.1.1 Main Applications of DSPs

A DSP is a special purpose processor that is different from a general purpose processor such as an Intel Pentium processor. While the latter is used for large memory, advanced operating applications, the DSP is a small, low-power consumption, low cost device. The sum of products (SOP) is the key element in DSP algorithms and is shown in Table 7.1.

### 7.1.2 Types and Sources of DSP Chips

Many companies produce DSP chips, including Analog Devices, Motorola, Lucent Technologies, NEC, SGS-Thompson, Conexant, and Texas Instruments.[1–3] In this laboratory, we will use DSP chips designed and manufactured

**TABLE 7.1**

Typical DSP Applications

| Algorithm | Equation |
|---|---|
| Finite impulse response filter | $y(n)=\sum_{k=0}^{M} a_k\, x(n-k)$ |
| Infinite impulse response filter | $y(n)=\sum_{k=0}^{M} a_k\, x(n-k) \quad + \sum_{k=1}^{N} b_k y(n-k)$ |
| Convolution | $y(n)=\sum_{k=0}^{N} x(k)h(n-k)$ |
| Discrete Fourier transform | $X(k)=\sum_{n=0}^{N-1} x(n)\exp[-j(2\pi / N)nk]$ |
| Discrete cosine transform | $F(u)=\sum_{x=0}^{N-1} c(u).f(x).\cos\left[\frac{\pi}{2N}u(2x+1)\right]$ |

by Texas Instruments (TI). These DSP chips will be interfaced through Code Composer Studio (CCS) software developed by TI.

#### *7.1.2.1 Evolution of Texas Instruments TMS320 DSP Chips*

In 1983, Texas Instruments (TI) released their first generation of DSP chips, the TMS320 single-chip DSP series. The first generation chips (C1x family) could execute an instruction in a single 200-nanosecond (ns) instruction cycle. The current generation of TI DSPs includes the C2000, C5000, and C6000 series, which can run up to eight 32-bit parallel instructions in one 6.67-ns instruction cycle, for an instruction rate of more than 1 GHz. The C2000 and C5000 series are fixed-point processors, while the C6000 series contains both fixed-point and floating-point processors. For this lab, we will be using the C6711 processor, the only C6000 series floating-point processor.[2]

The C2000 and C5000 series of chips are used primarily for digital control. They consume very little power and are used in many portable devices including 3G (third generation) cell phones, GPS (global positioning system) receivers, portable medical equipment, and digital music players. Due to their low power consumption (40 mW to 160 mW of active power), they are very attractive for power-sensitive portable systems. The C6000 series of chips provides both fixed- and floating-point processors that are used in systems that require high performance. Because these chips are not as power efficient as the C5000 series of chips (0.5 W to 1.4 W of active power), they are generally not used in portable devices. Instead, the C6000 series of chips is used in high quality digital audio applications, broadband infrastructure, and digital video imaging, the latter being associated almost exclusively with the fixed-point C64x family of processors. The key issues in DSP system design are power consumption, processing power, size, reliability, and efficiency.

In earlier times, assembly language was preferred for DSP programming. Today, C is the preferred way to code algorithms, and we shall use it for fixed- and floating-point processing. Please refer to Appendix E for more details on the TI 6000 series DSP.

### 7.1.3 TMS320C6711 DSP Starter Kit

The TMS320C6711 DSP chip is very powerful by itself, but for development of programs, a supporting architecture is required to store programs and data and to bring signals on and off the board. In order to use this DSP chip in a lab, a circuit board is provided that contains appropriate components. Together, Code Composer Studio (CCS), DSP chip, and supporting hardware make up the DSP Starter Kit, or DSK. A photograph of the TMS320C6711 DSK is shown in Figure 7.1.[4] The following hardware is included with the 'C6711 DSK board:

- 150 MHz 'C6711 DSP
- 16 MB External SDRAM and 128 KB External Flash; provides additional program and data storage
- TI's TLC320AD535 16-Bit Data Converter
- TI's TPS56100 Power Management Device
- JTAG Controller; provides easy emulation and debugging
- Expansion Daughter Card Interface; provides extensible system development
- CE-Compliant Universal Power Supply for DSK

### 7.1.4 Programming Languages

Assembly language was once the most commonly used programming language for DSP chips (such as TI's TMS320 series) and microprocessors (such as Motorola's 68MC11 series). Coding in assembly forces the programmer to manage CPU core registers (located on the DSP chip) and to schedule events in the CPU core. It is the most time consuming way to program, but it is the only way to fully optimize a program. Assembly language is specific to a given architecture and is primarily used to schedule time-critical and memory-critical parts of algorithms.

The preferred way to code algorithms is to code them in C. Coding in C requires a compiler that will convert C code to the assembly code of a given DSP instruction set. C compilers are very common, so this is not a limitation. In fact, it is an advantage, because C coded algorithms may be implemented

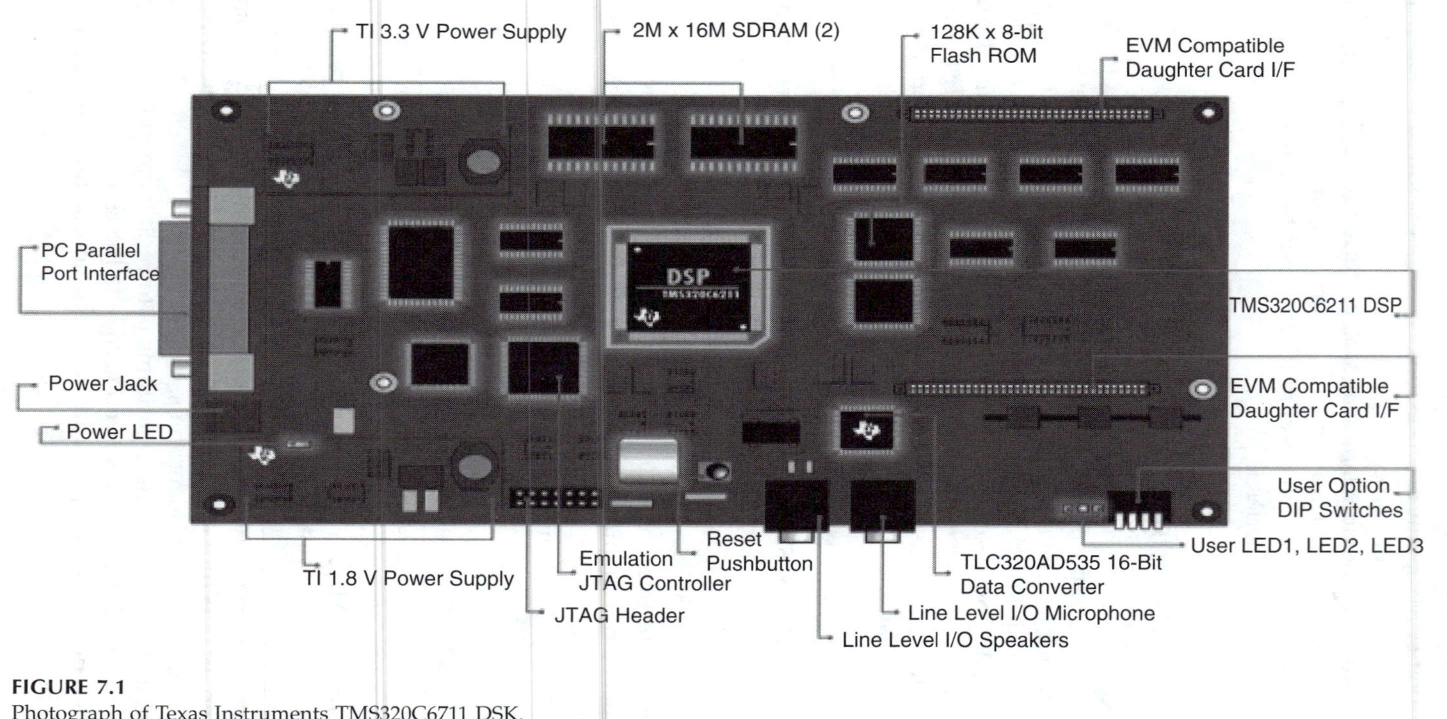


**FIGURE 7.1**
Photograph of Texas Instruments TMS320C6711 DSK.

on a variety of platforms (provided there is a C compiler for a given architecture and instruction set). In CCS, the C compiler has four optimization levels. The highest level of optimization does not achieve the same level of optimization that programmer-optimized assembly programs does, but TI has done a good job in making the optimized C compiler produce code that is comparable with programmer-optimized assembly code.

Finally, a hybrid between assembly language and C exists within CCS. It is called *linear assembly code*. Linear assembly looks much like assembly language code, but it allows for symbolic names and does not require the programmer to specify delay slots and CPU core registers on the DSP. Its advantage over C code is that it uses the DSP more efficiently, and its advantage over assembly code is that it does not require the programmer to manage the CPU core registers.

## 7.2 Software/Hardware Laboratory Using the TI TMS320C6711 DSK

### 7.2.1 Software and Hardware Equipment Requirements

For most exercises in this laboratory, the following equipment will be needed at every lab station:

- A Pentium-based computer with CCS version 2.0 or greater installed on it
- A 'C6711 DSK including power supply and parallel printer port cable
- Two coaxial cables with an 1/8-inch stereo headphone male jack on one end and two BNC male connectors (RF connectors) on the other end
- A set of speakers or headphones
- One coaxial cable with 1/8-inch stereo headphone jacks on both ends
- A signal generator
- An oscilloscope

### 7.2.2 Initial Setting Up of the Equipment

- Connect the parallel printer port cable between the parallel port on the DSK board (J2) and the parallel printer port on the back of the computer.
- Connect the 5V power supply to the power connector next to the parallel port on the DSK board (J4). You should see 3 LEDs blink next to some dip switches.

Once the DSK board is connected to your PC and the power supply has been connected, you can start CCS.

- To do this, click on **Start** on your computer, go to **Program**, and then go to **Texas Instruments**, then **Code Composer Studio DSK Tools 2 ('C6000)**, and select **CCStudio**.
- Or from **Desktop** click on **CCS-DSK 2 ('C6000)** icon

### 7.2.3 Study and Testing of the Code Composer Studio (CCS)

**Exercise 1: This first experiment in basic 'C6711 DSK commands consist of a series of twelve steps. Please follow the instructions carefully to successfully complete the experiment.**

The Code Composer Studio (CCS) is a powerful integrated development environment (IDE) that provides a useful transition between a high-level (C or assembly) DSP program and an on-board machine language program. CCS consists of a set of software tools and libraries for developing DSP programs, compiling them into machine code, and writing them into memory on the DSP chip and on-board external memory. It also contains diagnostic tools for analyzing and tracing algorithms as they are being implemented on-board.

This exercise will familiarize you with the software while covering the following key sections: Creating Projects, Debugging and Analysis, and Resets.

**Step 1: Creating a "Project"**

- In CCS, Choose **Project > New**.
- Type in a **Project Name** and a **location** where it will be stored.
- The type should be **.out** and the target **67xx.**
- Hit **Finish** key.
- Your **project name.pjt** should be in the **Project View** window on the lefthand side.

**Step 2: Creating a Source File**

- Choose **File > New > Source File**.
- An Editor window comes up.
- Type in the following assembly code, which declares 10 values:

```
.sect ".data"
.short 0
.short 7
```

```
.short 10
.short 7
.short 0
.short -7
.short -10
.short -7
.short 0
.short 7
```

- Choose **File > Save As**
- In your project folder, Save the file as **initializemem**, and choose type **.asm**.
- Create another source file, **main.c,** given below, and save in your project folder.

---

```
//main.c C program
#include <stdio.h>
void main()
{
      printf("Begin\n");
      printf("End\n");
}
```

---

- Currently, the program does nothing. However, it will be developed later.

**Step 3: Creating a Command File**

- Locate the **Hello.cmd** file on the computer.
- Choose **File** > **Open** and open the file.
- After the line: **Sections {**
- Type in **.data > SDRAM**.

This will put the data from your **initializemem** file in a part of **SDRAM** starting at address **0x80000000**.

- Choose **File** > **Save as**. Save the file in your project folder as **Lab1.cmd**.

Although the files you have created are in your project folder, they have not been put in the folders that will be used for assembling and linking. We have to add these files to the project for this purpose.

**Step 4: Adding Files to a Project**

- Select **Project > Add Files to Project**.
- Open the **initializemem.asm** file from your project folder. This file should now be under the **Source** folder in the **Project View** window.
- Repeat the above instruction for the **main.c** and **Lab1.cmd** files.

We must also add the run-time support library for the board, since we have a C program. Add the file located at **c:\ti\c6000\cgtools\lib\rts6701.lib**. This file should now appear under the **Libraries** folder in the **Project View** window.

**Step 5: Creating the executable file, lab1.out**

Before we compile, assemble, and link, there are a number of options we can choose to determine the amount of optimization to be done. There are four levels (Opt Levels) of optimization: 0, 1, 2, and 3. The lowest level is 0. However, sometimes, debugging cannot be done when we use optimization.

- Select **Project > Build Options**.
- Select **Compiler** and in the **Category** column, click on **Basic**. Check that **Target Version** is 671x and **Opt Level** is None.
- Similarly, select **Linker** and then **Basic**. You can change the name of the executable file that will be produced. Change the output file name to **Lab1.out**.
- **Project > Rebuild All** compiles, assembles, and links all of the files in the project and produces the executable file **lab1.out**. A window at the bottom shows if there are errors.
- **Project > Build** can be used when you have made a change to only a few files and now wish to compile, assemble, and link with the changed files. There are shortcut buttons on the window to do **Project Build** and **Rebuild**.

Upon building, there should have been a lot of errors. Scroll up until you reach the first red line with **error!** in it. Double click on the line. The file **initializemem.asm** opens at the line where the error occurred. Assembly code requires that all of the lines in the assembly file *not* start in the first column. So enter a space at the beginning of each line in the file and then save the file. Since we didn't change every file in the project, we can do a **Project > Build**.

**Step 6: Running the Program**

In order to run the program, we need to load the program into the DSP memory.

- Select **File > Load Program**.
- Open the **LAB1.out** program, which is in the **Debug** folder of your **LAB1** project folder.
- A **Disassembly** window should appear.
- To run select **Debug > Run**.

**Begin** and **End** should appear in the bottom **Stdout** window.

**Step 7: Viewing Memory**

This step is to check if the values of our **intializemem** file are in the memory location that we established in the **.cmd** file.

- Select **View** > **memory**.
- Type in **0x80000000** in **Address memory** location.
- Select **Format**: **16bit Signed Int**.
- A **Memory** window appears with the memory addresses and their contents.
- Compare the first 10 values with the **initializemem** file data.

**Step 8: Graphical Display of Data**

In order to view the graph of data in memory, complete the following instructions:

- **View > Graph > Time/frequency**.
- Set **Start Address: 0x80000000**.
- Set **Acquisition Buffer Size: 10**.
- Set **Display Data Size: 10**.
- Set **DSP Data Type: 16-bit signed integer**.
- A graph should appear on the screen with a plot of your data.
- Compare the first 10 values with the **initializemem** file data

**Step 9: Main.c Program Modification**

- Double click on **main.c** in the project window.
- Modify **main.c** program so that it looks like the following:

---

```
//main.c C program -Modification 1
#include <stdio.h>
void main()
{
```

```
    int i;
    short *point;
    point = (short *) 0x80000000;
    printf("Begin\n");
    for (i=0;i<10;i++)
    {
        printf("[%d]%d\n,"i, point[i]);
    }
    printf("End\n");
}
```

- **Save**, **Rebuild,** and **Load** the program into the DSP memory.
- **Run** the program.

A pointer is assigned to the beginning of our data in memory. This allows us to bring data into our c program and print out the data.

**Step 10: Checking a Variable during Program Execution**

**Breakpoints** and **watch windows** are used to watch variables while a program runs. In order to look at the values of the variable pointer in **main.c**, before and after the pointer assignment, as well as the value of variable **i**, first we establish breakpoints as follows:

- Select **File** > **Reload** to reload the program into DSP memory.
- Double click on **main.c** in the **project window**.
- Put a cursor on the line: **point = (short*) 0x80000000**.
- **Right click** and choose **Toggle Breakpoint**.
- Click on **Stdout window** in order to see the output results.
- Repeat the above procedure with the line: **printf("[%d]%d\n,"i, point[i]);**.

In order to add variables to **watch window**:

- Use the mouse to highlight the variable point in the line beginning with: **point = (short*)**.
- **Right click** and select **Add to Watch window**. A **watch window** should open with variable point.
- Repeat above procedure for variable **i** in the line beginning with: **printf("[%d]**.
- Select **Debug** > **Run**.

The program stops at the **breakpoint** and the **watch window** shows the value of point before the pointer is set. To advance the program or move through the **breakpoint**, hit the shortcut button **step over a line** or select **Debug** > **Step Over**. The pointer is now set, and you can see the value of the pointer is **0x80000000**.

To watch the variable **i** as the program progresses:

- Hit the shortcut button **animate** or select **Debug** > **animate** or hit the shortcut button **step over the line** over and over to see the variable **i** change.
- After using **animate**, you need to **halt** the system. You can do this with **Debug** > **Halt**.
- If you want to do this exercise over again, go to **Debug** > **Restart**, **Run**, **Step Over**, etc. Remove the **breakpoints** before continuing by hitting the shortcut button **Remove All Breakpoints**.
- Double click on **main.c** in the project window and modify the C program so that it matches the program below. This C program will sum the values.

---

```
//main.c C program -Modification 2
#include <stdio.h>
void main()
{
  int i, ret;
  short *point;
  point= (short *) 0x80000000;
  printf("Begin\n");
  for (i=0;i<10;i++)
  {
      printf("[%d]%d\n,"i, point[i]);
  }
  ret = ret_sum(point,10);
  printf("Sum =%d\n,"ret);
  printf("End\n");
}
int ret_sum(const short* array, int N)
{
  int count, sum;
  sum=0;
```

```
  for(count=0; count<N; count++)
      sum += array[count];
  return(sum);
}
```

- Go through all the steps required to run the program.

**Step 11: Benchmarking**

Now we will **benchmark** or time the subroutine to determine how long it takes to return the sum.

- **Reload** the program.
- Select **Profiler** > **Start New Session**.
- Title the session **Lab 1**. A **profile** window comes up in the bottom.
- Double click on **main.c** in the project window.
- Put your cursor on the line: **int ret_sum(const short* array, int N)**.
- Several shortcut buttons are on the left side of the **Profile window**.
- Hit the **Create Profile Area** button. Make sure the type is **Function** and the line number corresponds to the beginning of the function, since this is where you placed the cursor.
- Hit **OK**.
- Expand **Lab1.out** under the Files window pane. The function **ret_sum** should be there.
- **Run** the program.
- The value for the **Incl. Total** in the profiler window is the number of clock cycles needed to run the function **ret_sum**.
- To redo this exercise, highlight **ret_sum** in the **Files window pane**, **right click**, and select **Clear Selected**.
- Then hit **Debug** > **Restart** and **Run** the program.

Optimization can change the amount of time required to run the function. To observe the effects, follow these instructions:

- Select **Project > Build Options**.
- Choose **Compiler, Basic,** and **Opt Level o0**.
- Select **Project > Rebuild All**.
- Select **File > Load Program Lab1.out**.
- Highlight **ret_sum** in the **Files pane**, **right click**, and select **Clear Selected**.
- Hit **Debug > Run**.

Repeat the above for the other levels of optimization **o1**, **o2**, and **o3**, and compare the number of clock cycles for each optimization.

**Step 12: Resets**

Sometimes it might be necessary to reset the DSK. There are several ways to do this at different levels.

- In **CCS** go to **Debug** > **Reset CPU** — A **Disassembly** window appears.
- Close **CCS**.
- Select **Start** button of Windows > **Programs** > **Texas Instruments** > **Code Composer Studio DSK Tools** > **Hardware Resets** > **Reset**.
- Close **CCS**. Unplug the board and turn it back on after a few seconds. Wait for the LEDs to quit flashing before trying to use the board or opening CCS again.
- Press the **Reset** button on the DSK. *Note:* Do not press this button while CCS is still running. Also, this reset does not perform the full reset.

### 7.2.4 Experimenting with the 'C6711 DSK as a Signal Source

**Exercise 2: This experiment tests the 'C6711 DSK as a signal source, and consists of seven steps. Please follow the instructions carefully to complete the experiment.**

**Step 1: Creating the Project File Sine_gen.pjt**

- In **CCS**, select **Project** and then **New**. A window named **Project Creation** will appear, as shown in Figure 7.2.
- In the field labeled **Project Name**, enter **Sine_gen**. In the field **Location**, click on the right side of the field and browse to the folder **c:\ti\myprojects\Sine_gen**.
- In the field **Project Type**, verify that **Executable (.out)** is selected, and in the field **Target**, verify that **TMS320C67XX** is selected.
- Finally, click on **Finish**. **CCS** has now created a project file **Sine_gen.pjt**, which will be used to build an executable program. This file is stored in the folder **c:\ti\myprojects\Sine_gen.** The **.pjt** file stores project information on build options, source filenames, and dependencies.

**Step 2: Creating Support Files**

- **C6xdsk.cmd**
    - Find and open **Hello.cmd** file in **c:\ti** directory and modify it to the following program given below.
    - Save it as **C6xdsk.cmd**, as shown below.

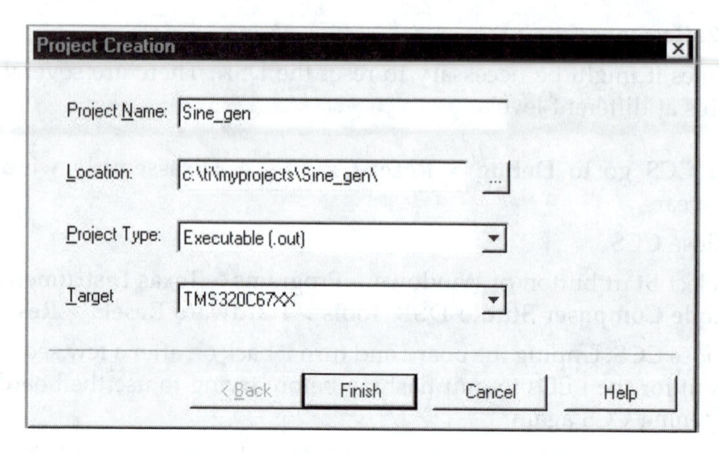


**FIGURE 7.2**
Project creation window for Sine_gen.pjt.

---

```
//*C6xdsk.cmd Generic Linker command file*/

MEMORY
{
 VECS:      org = 0h, len = 0x220
 IRAM:      org = 0x00000220, len = 0x0000FDC0/
            *internal memory*/
 SDRAM:     org = 0x80000000, len = 0x01000000/
            *external memory*/
 FLASH:     org = 0x90000000, len = 0x00020000/*flash
            memory*/
}

SECTIONS
{
 vectors   :> VECS
 .text     :> IRAM
 .bss      :> IRAM
 .cinit    :> IRAM
 .stack    :> IRAM
 .sysmem   :> SDRAM
 .const    :> IRAM
```

```
 .switch   :> IRAM
 .far      :> SDRAM
 .cio      :> SDRAM
}
```

- **C6xdskinit.c**
  - Create a file with the following listing and save it as **C6xdskinit.c,** as shown below.
  - To create file go to **File> new> source file**.

```
//C6xdskinit.c Init DSK,AD535,McBSP (includes functions
provided with DSK)
#include <c6x.h>
#include "c6xdsk.h"
#include "c6xdskinit.h"
#include "c6xinterrupts.h"

char polling = 0;

void mcbsp0_init()                //set up McBSP0
{
 *(unsigned volatile int *)McBSP0_SPCR = 0; //reset
serial port
 *(unsigned volatile int *)McBSP0_PCR = 0; //set pin
control reg
 *(unsigned volatile int *)McBSP0_RCR = 0x10040;//set rx
control reg one 16 bit data/frame
 *(unsigned volatile int *)McBSP0_XCR = 0x10040;//set tx
control reg one 16 bit data/frame
 *(unsigned volatile int *)McBSP0_DXR = 0;
 *(unsigned volatile int *)McBSP0_SPCR = 0x12001;//setup
SP control reg
}

void mcbsp0_write(int out_data)    //function for writing
{
int temp;
```

```
if (polling)      //bypass if interrupt-driven
{
 temp = *(unsigned volatile int *)McBSP0_SPCR & 0x20000;
 while (temp == 0)
  temp = *(unsigned volatile int *)McBSP0_SPCR & 0x20000;
}
*(unsigned volatile int *)McBSP0_DXR = out_data;
}

int mcbsp0_read()     //function for reading
{
int temp;

if (polling)
{
temp = *(unsigned volatile int *)McBSP0_SPCR & 0x2;
while (temp == 0)
 temp = *(unsigned volatile int *)McBSP0_SPCR & 0x2;
}
temp = *(unsigned volatile int *)McBSP0_DRR;
return temp;
}

void TLC320AD535_Init()   //init AD535
{
 mcbsp0_read(); //setting up AD535 Register 3
 mcbsp0_write(0);
 mcbsp0_read();
 mcbsp0_write(0);
 mcbsp0_read();
 mcbsp0_write(0);
 mcbsp0_read();
 mcbsp0_write(1); //send bit for Secondary Communications
 mcbsp0_read();
 mcbsp0_write(0x0386); //voice channel reset, pre-amps
selected
 mcbsp0_read();
 mcbsp0_write(0); //clear Secondary Communications
```

```
 mcbsp0_read();
 mcbsp0_write(0);
 mcbsp0_read();
 mcbsp0_write(0);
 mcbsp0_read();
 mcbsp0_write(1); //send bit for Secondary Communications
 mcbsp0_read();
 mcbsp0_write(0x0306); //voice channel clear reset, pre-
amps selected
 mcbsp0_read();
 mcbsp0_write(0); //clear Secondary Communications
 mcbsp0_read();
 mcbsp0_write(0); //setting up AD535 Register 4
 mcbsp0_read();
 mcbsp0_write(0);
 mcbsp0_read();
 mcbsp0_write(1);
 mcbsp0_read();
 mcbsp0_write(0x0400); //set microphone pre-amp gain to
20 dB
 mcbsp0_read();
 mcbsp0_write(0);
 mcbsp0_read();
 mcbsp0_write(0); //setting up AD535 Register 5
 mcbsp0_read();
 mcbsp0_write(0);
 mcbsp0_read();
 mcbsp0_write(1);
 mcbsp0_read();
 mcbsp0_write(0x0502); //DAC PGA = 0 dB
 mcbsp0_read();
 mcbsp0_write(0);
 mcbsp0_read();
}

void c6x_dsk_init()    //dsp and peripheral init
{
```

```
CSR=0x100;              //disable all interrupts
IER=1;            //disable interrupts except NMI
ICR=0xffff; //clear pending interrupts
*(unsigned volatile int *)EMIF_GCR = 0x3300; //EMIF
global control
*(unsigned volatile int *)EMIF_CE0 = 0x30; //EMIF CE0
control
*(unsigned volatile int *)EMIF_CE1 = 0xffffff03;
//EMIF CE1 control,8bit async
*(unsigned volatile int *)EMIF_SDCTRL = 0x07117000;//EMIF
SDRAM control
*(unsigned volatile int *)EMIF_SDRP = 0x61a; //EMIF SDRM
refresh period
*(unsigned volatile int *)EMIF_SDEXT = 0x54519;
//EMIF SDRAM extension
mcbsp0_init();
TLC320AD535_Init();
}

void comm_poll()  //for communication/init using polling
{
 polling = 1;      //if polling
 c6x_dsk_init();  //call init DSK function
}

void comm_intr()  //for communication/init using
interrupt
{
 polling = 0;      //if interrupt-driven
 c6x_dsk_init();  //call init DSK function
 config_Interrupt_Selector(11, XINT0);//using transmit
interrupt INT11
 enableSpecificINT(11); //for specific interrupt
 enableNMI(); //enable NMI
 enableGlobalINT();//enable GIE for global interrupt
 mcbsp0_write(0); //write to SP0
}

void output_sample(int out_data) //added for output
{
```

```
 mcbsp0_write(out_data & 0xfffe); //mask out LSB
}

int input_sample() //added for input
{
 return mcbsp0_read();//read from McBSP0
}
```

- **Vectors_11.asm**

  Locate the **Vectors.asm** file in **c:\ti** folder. Modify **Vectors.asm** as below to handle interrupts. Save it as **Vectors_11.asm,** as shown below. Twelve interrupts, **INT4** through **INT15**, are available, and **INT11** is selected within this vector file.

```
//*Vectors_11.asm Vector file for interrupt-driven program
           .ref             _c_int11; ISR used in C program
           .ref  _c_int00  ;entry address
           .sect "vectors" ;section for vectors
RESET_RST: mvkl  .S2        _c_int00,B0;lower 16 bits -> B0
           mvkh .S2_c_int00,B0 ;upper 16 bits -> B0
           B     .S2  B0      ;branch to entry address
           NOP                ;NOPs for remainder of FP
           NOP                ;to fill 0x20 Bytes
           NOP
           NOP
           NOP
NMI_RST:   .loop 8
           NOP                ;fill with 8 NOPs
           .endloop
RESV1:     .loop 8
           NOP
           .endloop
RESV2:     .loop 8
           NOP
           .endloop
INT4: .loop 8
           NOP
           .endloop
```

```
INT5: .loop 8
          NOP
          .endloop
INT6: .loop 8
          NOP
          .endloop
INT7:     .loop 8
          NOP
          .endloop
INT8: .loop 8
          NOP
          .endloop
INT9:           .loop 8
          NOP
          .endloop
INT10: .loop 8
          NOP
          .endloop

INT11:    b    _c_int11     ;branch to ISR
          .loop 7
          NOP
          .endloop

INT12:    .loop 8
          NOP
          .endloop
INT13: .loop 8
          NOP
          .endloop
INT14:    .loop 8
          NOP
          .endloop
INT15:    .loop 8
          NOP
          .endloop
```

- **C6xdsk.h**
  - Find file **C6211dsk.h** in the **C:\ti** directory.
  - Save this file as **c6xdsk.h**.
- **C6xinterrupts.h**
  - Create a file with the following listing and save it as **C6xinterrupts.h**.

---

```
//C6xinterrupts.h provided by TI
#define DSPINT    0x0 /* 00000b DSPINT Host port host to
                         DSP interrupt */
#define TINT0     0x1 /* 00001b TINT0 Timer 0 interrupt */
#define TINT1     0x2 /* 00010b TINT1 Timer 1 interrupt */
#define SD_INT    0x3 /* 00011b SD_INT EMIF SDRAM timer
                         interrupt */
#define EXT_INT4 0x4 /* 00100b EXT_INT4 External
                         interrupt 4 */
#define EXT_INT5 0x5 /* 00101b EXT_INT5 External
                         interrupt 5 */
#define EXT_INT6 0x6 /* 00110b EXT_INT6 External
                         interrupt 6 */
#define EXT_INT7 0x7 /* 00111b EXT_INT7 External
                         interrupt 7 */
#define EDMA_INT 0x8 /* 01000b EDMA_INT EDMA channel
                         (0 through 15) interrupt */
#define XINT0     0xC /* 01100b XINT0 McBSP 0 transmit
                         interrupt */
#define RINT0     0xD /* 01101b RINT0 McBSP 0 receive
                         interrupt */
#define XINT1     0xE /* 01110b XINT1 McBSP 1 transmit
                         interrupt */
#define RINT1     0xF /* 01111b RINT1 McBSP 1 receive
                         interrupt */

/*******************************************************
* Interrupt Initialization Functions
*
* (CSR and IER are CPU registers defined in c6x.h)
*
*******************************************************
  */
```

```
/* Enable Interrupts Globally (set GIE bit in CSR = 1) */
void enableGlobalINT(void)
{
 CSR |= 0x1;
}

/* Enable NMI (non-maskable interrupt); must be enabled
* or no other interrupts can be recognized by 'C6000 CPU */
void enableNMI(void)
{
 IER = _set(IER, 1, 1);
}

/* Enable a specific interrupt;
 * (INTnumber = {4,5,6, …,15}) */
void enableSpecificINT(int INTnumber)
{
 IER = _set(IER, INTnumber, INTnumber);
}

/*******************************************************
* C6000 devices have hardware configurable interrupts.
* To use the McBSP interrupts you must configure them
  because they are selected by default.
* You must set the appropriate interrupt select bits in
  IML and IMH memory-mapped int select registers.
* IML and IMH addresses are defined in c6211dsk.h.
*******************************************************/
void config_Interrupt_Selector(int INTnumber, int
  INTsource)
{
 /* INTnumber = {4,5,6, …,15}
    INTsource = see #define list above
*/

union
{
struct
```

```
{ unsigned int INTSEL4 : 5;
  unsigned int INTSEL5 : 5;
  unsigned int INTSEL6 : 5;
  unsigned int rsvbit15 : 1;
  unsigned int INTSEL7 : 5;
  unsigned int INTSEL8 : 5;
  unsigned int INTSEL9 : 5;
  unsigned int rsvbit31 : 1;
}exp;
unsigned int reg;
}IMLvalue;                          /* = {0,0,0,0,0,0,0,0}; */

union
{
unsigned int reg;
struct
{ unsigned int INTSEL10 : 5;
unsigned int INTSEL11 : 5;
  unsigned int INTSEL12 : 5;
  unsigned int rsvbit15 : 1;
  unsigned int INTSEL13 : 5;
unsigned int INTSEL14 : 5;
  unsigned int INTSEL15 : 5;
  unsigned int rsvbit31 : 1;
}exp;
}IMHvalue;

IMLvalue.reg = *(unsigned volatile int *)IML;
IMHvalue.reg = *(unsigned volatile int *)IMH;

switch (INTnumber)
{
 case 4 :
  IMLvalue.exp.INTSEL4 = INTsource;
  break;
```

```
case 5 :
 IMLvalue.exp.INTSEL5 = INTsource;
 break;
case 6 :
 IMLvalue.exp.INTSEL6 = INTsource;
 break;
case 7 :
 IMLvalue.exp.INTSEL7 = INTsource;
 break;
case 8 :
 IMLvalue.exp.INTSEL8 = INTsource;
 break;
case 9 :
 IMLvalue.exp.INTSEL9 = INTsource;
 break;
case 10 :
 IMHvalue.exp.INTSEL10 = INTsource;
 break;
case 11 :
 IMHvalue.exp.INTSEL11 = INTsource;
 break;
case 12 :
 IMHvalue.exp.INTSEL12 = INTsource;
 break;
case 13 :
 IMHvalue.exp.INTSEL13 = INTsource;
 break;
case 14 :
 IMHvalue.exp.INTSEL14 = INTsource;
 break;
case 15 :
 IMHvalue.exp.INTSEL15 = INTsource;
 break;
```

```
 default : break;
}

*(unsigned volatile int *)IML = IMLvalue.reg;
*(unsigned volatile int *)IMH = IMHvalue.reg;
return;
}
```

- **C6xdskinit.h**
  Create a file with the following listing and save it as **C6xdskinit.h**.

```
//C6xdskinit.h Function prototypes for routines in
  c6xdskinit.c
void mcbsp0_init();
void mcbsp0_write(int);
int mcbsp0_read();
void TLC320AD535_Init();
void c6x_dsk_init();
void comm_poll();
void comm_intr();
int input_sample();
void output_sample(int);
```

- **Amplitude.gel**
  Create a file with the following listing and save it as **amplitude.gel**.

```
/*Amplitude.gel Create slider and vary amplitude of
sinewave*/

menuitem "Sine Amplitude"

slider Amplitude(10000,35000,5,1,amplitudeparameter)/
*incr by 5,up to 35000*/
{
 amplitude = amplitudeparameter; /*vary amplitude of sine*/
}
```

**Step 3: Adding Support Files to a Project**

The next step in creating a project is to add the appropriate support files to the file **Sine_gen.pjt**.

- In the **CCS window**, go to **Project** and then select **Add Files to Project**.
- In the window that appears, click on the folder next to where it says **Look In:** Browse to folder where you stored the support files created earlier. You should be able to see the file **C6xdskinit.c**. Notice that the **Files of type** field is **C source code.**
- Click on **C6xdskinit.c** and then click on **Open**.
- Repeat this process two more times, adding the files **vectors_11.asm** and **C6xdsk.cmd** to the project file **Sine_gen.pjt**.
- For field, select **Files of type**, select **asm Source Files (*.a*)**.
- Click on **vectors_11.asm** and then click on **Open**.
- For field, select **Files of type**, select **Linker Command File (*.cmd)**.
- Click on **C6xdsk.cmd** and then click on **Open**. You have now created your project file **c:\ti\myprojects\Sine_ gen.pjt**.

The C source code file contains functions for initializing the DSP and peripherals. The **vectors** file contains information about what interrupts (if any) will be used and gives the linker information about resetting the CPU. This file needs to appear in the first block of program memory. The linker command file (**C6xdsk.cmd**) tells the linker how the **vectors** file and the internal, external, and flash memory are to be organized in memory.

In addition, it specifies what parts of the program are to be stored in internal memory and what parts are to be stored in the external memory. In general, the program instructions and local and global variables will be stored in internal (random access) memory or IRAM.

**Step 4: Adding Appropriate Libraries to a Project**

In addition to the support files that you have been given, there are precompiled files from TI that need to be included with your project. For this project, you need a run-time support library (**rts6701.lib**), which your support files will use to run the DSK, and a **gel** (general extension language) file (**dsk6211_6711.gel**) to initialize the DSK. The **gel** file was automatically included when the project file **Sine_gen.pjt** was created, but the **RTS** (run-time support) library must be included in the same manner used to include the previous files.

- Go to **Project** and then select **Add Files to Project**.
- For **Files of type**, select **Object and Library Files (*.o*,*.l*)**.

- Browse to the folder **c:\ti\c6000\cgtools\lib** and select the file **rts6701.lib** (which supports the C67x/C62x architecture).
- In the left subwindow of the **CCS** main window, double click on the folder **Libraries** to make sure the file was added correctly.

These files, along with your other support files, form the *black box* that will be required for every project created in this lab. The only files that change are the source code files that code a DSP algorithm and possibly a vectors file.

**Step 5: Adding Source Code Files to a Project**

The last file that you need to add to **Sine_gen.pjt** is your source code file. This file will contain the algorithm that will be used to internally generate a 1KHz sine wave.

- Create a file with the following code. Save it as **Sine_gen.c**.

```
//Sine_gen.c C program file to generate sine wave
#include <math.h>            //needed for sin() function
#define PI 3.14159265359     //define the constant PI
float f0=1000;               //sinusoid frequency
short fs=8000;               //sampling frequency of codec
float angle=0;               //angle in radians
float offset;                //offset value for next sample
short sine_value;            //value sent to codec
short amplitude = 20000;     //gain factor

interrupt void c_int11()     //interrupt service routine
{
offset=2*PI*f0/fs;           //set offset value
angle = angle + offset;      //previous angle plus offset

if (angle > 2*PI)            //reset angle if > 2*PI
angle -= 2*PI;               //angle = angle - 2*PI

sine_value=(short)amplitude*sin(angle); //calculate
                                    current output sample
output_sample(sine_value);//output each sine value
return;                      //return from interrupt
}
```

```
void main()
{
comm_intr();        //init DSK, codec, SP0 for interrupts
while(1);           //wait for an interrupt to occur
}
```

- Go back to **Project** and then **Add Files to Project**.
- Select the file **Sine_gen.c** and add it to your project by clicking on **Open**.
- You may have noticed that the **.h** files cannot be added. These files are header files and are referenced in **C6xdskinit.c**.
- Go to **Project** and select **Scan All Dependencies**.
- In **CCS**, double click on **Sine_gen.pjt** and then double click on **Include**. You should see the three header files that you added plus a mystery file, **C6x.h**. This mystery file is included with the Code Composer Studio software, and it is used to configure the board.
- Open the file **C6xdskinit.c** and observe that the first four lines of code include the four header files. The project file **Sine_gen.pjt** has now been charged with all of the files required to build the executable.out file.

**Step 6: Build Options**

The next objective is to customize the compiler and linker options so the executable file gets built correctly. Also, the compiler will first convert the C coded programs into DSP assembly programs before it compiles them into machine code. By selecting the appropriate options, we can keep these intermediate assembly files. For your own amusement, you can open these files in a word processing program to see how the DSP assembly is coded. To make these customizations:

- Click on the **Project** pull-down menu; go to **Build Options**. This will open a new window, as shown in Figure 7.3.
- In this window, click on the **Compiler** tab.
- In the **Category** column, click on **Basic** and select the following:
    - **Target Version:** 671x
    - **Generate Debug Info:** Full Symbolic Debug (-g)
    - **Opt Speed vs. Size:** Speed Most Critical (no ms)
    - **Opt Level:** None
    - **Program Level Opt:** None

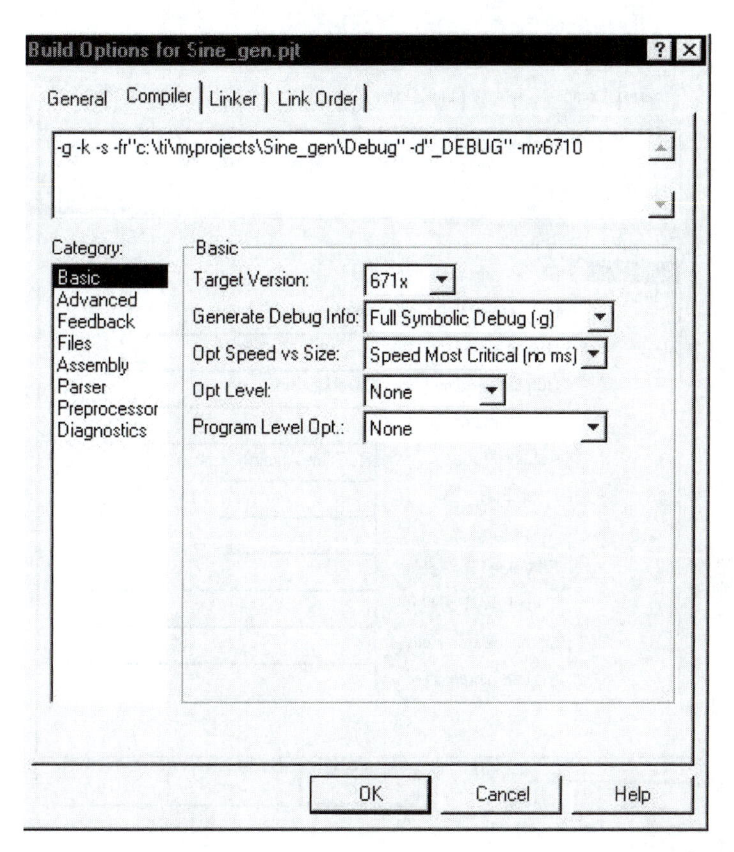


**FIGURE 7.3**
Build options for compiling.

- In the top part of the current window, you should see:

  `-g -q -fr"c:\ti\myprojects\Sine_gen\Debug" -d"_DEBUG" -mv6710`

- Change it to:

  `-g -k -s -fr" c:\ti\myprojects\Sine_gen\Debug " -d" DEBUG" -mv6710`

- Now click on the **Linker** tab on the top of the current window and make sure the following command appears in the top-most window (see Figure 7.4):

  `-q -c -o."\Debug\Sine_gen.out" -x`

  The options -g, -k, -s in the compiler options and -g, -c, -o in the linker options do serve a purpose,[4] but we will not be concerned with them just yet. Your project has now been created. This process is cumbersome, but it needs to be done only once. In future projects,

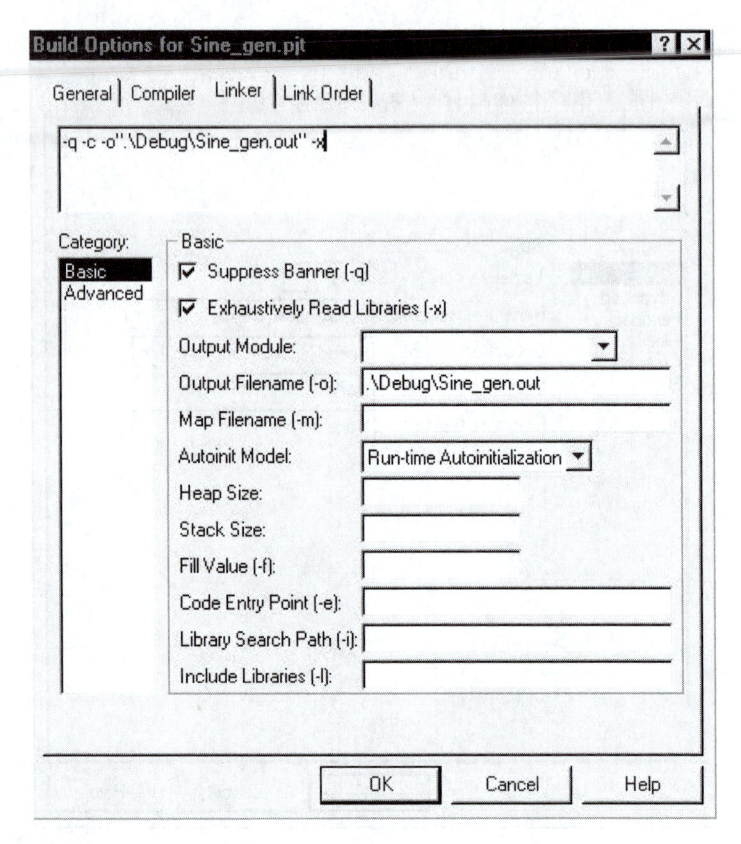


**FIGURE 7.4**
Build options for linking.

you will be able to copy this folder into another folder and make a few simple modifications. These modifications include altering the C code in **Sine_gen.c** and editing one linker option.

### Step 7: Building and Running the Project

Now you must build and run the project. To build the first project:

- Go to **Project** pull-down menu in the CCS window, and then select **Build** (or press the button with three red down arrows on the top toolbar in the CCS window). A new subwindow will appear on the bottom of the CCS window.

  When building is complete, you should see the following message in the new subwindow:

```
Build Complete,
0 Errors, 0 Warnings, 0 Remarks.
```

When CCS "built" your project, it compiled the C-coded source files and header files into assembly code, using a built-in compiler. Then it assembled the assembly code into a **COFF** (common object file format) file that contains the program instructions, organized into modules. Finally, the linker organized these modules and the run-time support library (**rts6701.lib**) into memory locations to create an **executable.out** file. This executable file can be downloaded onto the DSK. When this executable file is loaded onto the DSK, the assembled program instructions, global variables, and run-time support libraries are loaded to their linker-specified memory locations.

At this point, you should have the following files established on your c: drive:

- C6xdsk.h, C6xdskinit.c, C6xdskinit.h, c6xinterrupts.h, vectors.asm, vectors_11.asm, C6xdsk.cmd, rts6701.lib, Amplitude.gel.
- c:\ti\myprojects\Sine_gen {Sine_ gen.pjt, Sine_ gen.c}
- c:\ti\myprojects\Sine_gen\Debug\Sine_gen.out

To test this newly built program **Sine_gen.out** on the DSK, you must first load the program onto the board. But, before a new program is loaded onto the board, it is good practice to reset the CPU. To reset the CPU:

- Click on the **Debug** pull-down menu and select **Reset CPU**.
- Then, to load the program onto the DSK, click on the **File** pull-down menu and select **Load Program**.
- In the new window that appears, double click on the folder **Debug**, select the **Sine_gen.out** file, and click on **Open**. This will download the executable file **Sine_gen.out** onto the DSK. A new window will appear within CCS entitled **Disassembly**, which contains the assembled version of your program. Ignore this window for now.

Before you run this program, make sure that the cable between the 1/8-inch headphone jack on the DSK board (the J6 connector) and the oscilloscope is connected, and make sure that the oscilloscope is turned on.

- In CCS, select the **Debug** pull-down menu and then select **Run**, or just simply click on the top **running man** on the left side toolbar.
- Verify a 1 kHz sine wave of amplitude approximately 1.75 volts peak to peak on the oscilloscope. Once you have verified the signal, disconnect the oscilloscope from the DSK and attach a pair of speakers or headphones to the DSK. You should hear a 1 kHz *pure tone*.
- After you have completed both of these tasks, either click on the icon of the blue **running man** with a red **X** on it or go to the **Debug** pull-down menu to select **Halt**.

### 7.2.5 Experimenting with the 'C6711 DSK as a Real-Time Signal Source

**Exercise 3: This experiment tests the 'C6711 DSK as a real-time signal source and consists of a series of four steps. Please follow the instructions carefully to complete the experiment.**

In many of the communication systems that we design, we want to be able to generate a sinusoid with arbitrary frequency $fo$. In Exercise 2, we generated the sinusoid $x(t) = \sin(2\pi fot)$, where $fo$ = 1kHz. In real-time digital systems, this requires *samples* of the signal $x(t)$ to be sent to the codec at a fixed rate. In the case of the on-board codec, samples are being sent at rate $fs$ = 8 kHz ($Ts$ = 0.125ms).

In C code, we generate samples of $x(t)$, namely $x[n] = x(nts) = \sin(2\pi nfo/fs)$, where $fs = 1/ts$, which is defined only for integer values of $n$. Here, the argument of the sine function $\theta[n] = 2\pi nfo/fs$ is a linear function that can be easily updated at each sample point. Specifically, at the time instance $n + 1$, the argument becomes

$$\theta[n+1] = 2\pi(n+1)fo/fs = \theta[n] + 2\pi fo/fs \tag{7.1}$$

which is the previous argument plus the offset $2\pi fo/fs$. This makes it possible to generate any sinusoid whose frequency is $fo < 3.6$ kHz. You may have expected the maximum frequency to be $fs = 2 = 4$kHz, but the codec requires oversampling.

**Step 1: Code Analysis and Modification**

In this step, we analyze the source code in **Sine_gen.c** to see exactly how this 1 kHz sine wave was generated. Note that in C (or more precisely C++) that text following // on any line is regarded as a comment and is ignored, when the program is compiled. A listing of **Sine_gen.c** is given below.

---

```
//Sine_gen.c C program file to generate sine wave
#include <math.h>               //needed for sin() function
#define PI=3.14159265359        //define the constant PI
float f0=1000;                  //sinusoid frequency
short fs=8000;                  //sampling frequency of codec
float angle=0;                  //angle in radians
float offset;                   //offset value for next sample
short sine_value;               //value sent to codec
short amplitude = 20000;        //gain factor
interrupt void c_int11()        //interrupt service routine
```

```
{
offset = 2*PI*f0/fs;          //set offset value
angle = angle + offset;       //previous angle plus offset

if (angle > 2*PI)             //reset angle if > 2*PI
angle -= 2*PI;                //angle = angle - 2*PI

sine_value=(short)amplitude*sin(angle); //calculate
                                current output sample
output_sample(sine_value); //output each sine value
return;                       //return from interrupt
}

void main()
{
comm_intr();                  //init DSK, codec, SP0 for
                                interrupts
while(1);                     //wait for an interrupt to
                                occur
}
```

In order to efficiently analyze the code, we will break it up into three sections, namely *section one* (lines 1 through 8), *section two* (lines 9 through 20), and *section three* (lines 22 through 26). Generally, the section containing the **main()** function, section three in this case, will always come last. In C, the function **main()** is always the starting point of the program. The linker knows to look for this function to begin execution. Therefore, a C program without a **main()** function is meaningless.

**Step 2: Analyzing the Code**

The first section of code (lines 1 through 8) is used for preprocessor directives and the definition of global variables. In C, the # sign signifies a preprocessor directive. In this lab, we will primarily use only two preprocessor directives, namely **#include** and **#define**. In line 1, the preprocessor directive, **#include <math.h>**, tells the preprocessor to insert the code stored in the header file math.h into the first lines of the code **sine gen.c** before the compiler compiles the file. Including this header file allows us to call mathematical functions such as sin(¢), cos(¢), tan(¢), etc. as well as functions for logarithms, exponentials, and hyperbolic functions. This header file is required for the sin(¢) function line 17. To see a full list of functions available with **math.h**, use the help menu in CCS. The next preprocessor directive defines the fixed point number PI, which approximates the irrational number π. Before compiling, the preprocessor will replace every occurrence of PI in **Sine_gen.c** with the number specified.

The next six lines (6 through 8) define the global variables: f0, fs, angle, offset, sine_value, and amplitude. The variables fs, sine_value, and amplitude are of type **short**, which means they hold 16-bit signed integer values. The variables f0, angle, and offset are of type float, which means they hold IEEE single precision (32-bit) floating-point numbers.[3] Notice that all of the lines that contain statements end with a semicolon. This is standard in C code. The only lines that do not get semicolons are function names, such as **c_int11()**, conditional statements such as **if()**, and opening and closing braces ({ }) associated with them.

The last section of code (lines 22 through 26) contains the function **main()**. The format of the **main()** function will *not* change from program to program. Lines 22, 23, and 26 will always be the first two lines and last line, respectively of this routine. The first line in **main()** (line 24) calls the function **comm intr()**. This function is located within the file **C6xdskinit.c**, which is one of the *support* files given to you. This function initializes the on-board codec, specifies that the transmit interrupt XINT0 will occur in SP0, initializes the interrupt INT11 to handle this interrupt, and allows interrupts INT4 through INT15 to be recognized by the DSP chip. To learn more about configuring the DSP chip for handling interrupts, examine the code in **C6xdskinit.c** and refer to References 3–6 in this chapter. Now, the DSP chip and codec have been configured to communicate via interrupts, which the codec will generate every 0.125ms. The program **Sine_gen.c** now waits for an interrupt from the codec, so an infinite loop keeps the processor idle until an interrupt occurs. This does not have to be the case, since an interrupt will halt the CPU regardless of whether it is processing or idling. But in this program, there is no other processing, so we must keep the processor idling while waiting for an interrupt.

The middle section of code (lines 9 through 20) is used to define the **Interrupt Service Routine** or **ISR**. When an interrupt occurs, the program branches to the ISR **c_int11()** as specified by **vectors 11.asm**. This interrupt generates the current sample of the sinusoid and outputs it to the codec. Line 11 determines the offset value 2π *fo/fs*. For a given *fo*, this value will not change, so it does not need to be calculated every time an interrupt occurs. However, by calculating this value here, we will be able to change the value of our sinusoid using a **Watch Window**. This is demonstrated in the next section. Line 12 calculates the current sample point by taking the value stored in the global variable angle and adding the offset value to it. The angle variable is, of course, the angle (in radians) that is passed to the sine function. NB: In C the command **angle += offset**; is shorthand for the command **angle = angle + offset;**. The *sin*($x$) function in C approximates the value of sin($x$) for any value of $x$, but a better and more efficient approximation will be computed if $0 \leq n \leq 2\pi$. Therefore, lines 14 and 15 are used to reset the value of sample if it is greater than $2\pi$. Because sin($x$) is periodic $2\pi$ in $x$, subtracting $2\pi$. from $x$ will not change the value of the output. Line 17 calculates the sine value at the current sample point. The value is *typecast* as (short) before it is stored in the variable sine value. Typecasting tells the compiler to convert a value from one data type to another before storing it in a variable or sending it to a function. In this case, the value returned from the sin() is a single precision floating-point number (between –1.0 and 1.0) that gets scaled by amplitude value (20000). By typecasting this number as a short (16-bit signed integer between the values –32768 and 32767), the CPU will round the number to the nearest integer and store it in a 16-bit signed integer format (2's complement). This value is scaled by 20000 for two reasons. First, it is needed so that rounding errors are minimized, and second, it amplifies the signal so it can be observed on the oscilloscope and heard through speakers or headphones. This scaling factor must be less than 32768 to prevent overdriving the codec. Line 18 sends the current **sine_value** to the codec by calling the function **output_sample()**. The code for **output_sample()** is located in file **C6xdskinit.c**. Open the file **C6xdskinit.c** in CCS and examine the code for this function. This function **output_sample()** forces the least significant digit of the sample that it receives to zero and sends it to a function **mcbsp0_write()**, which writes the sample to the transmit buffer in the McBSP. This will cause the McBSP to transmit the data sample to the on-board codec. The masking of the least significant digit of the output sample is needed so that the on-board codec interprets the received binary number as a data sample and not as secondary information. Upon completion of the interrupt (generating a sinusoid sample and outputting it to the on-board codec), the interrupt service routine restores the saved execution state (see the command **return;** in line 19). In this program, the saved execution state will always be the infinite while loop in the **main()** function.

**Step 3: Using a Watch Window**

Once an algorithm has been coded, it is good to have software tools for observing and modifying the local and global variables after a program has been loaded onto the DSK. Located in CCS is a software tool called a **Watch Window**, which allows the user to view local variables and to view and modify global variables during execution. In this lab, we will not view any local variables, but we will view and modify global variables.

- In CCS, start running the program **sine_gen.out** again and make sure that you have a valid output on an oscilloscope.
- Then click on the pull-down menu **view**, and select **Watch Window**. A subwindow should appear on the bottom of the main CCS window. You should notice two tabs on the bottom left part of the new subwindow: **Watch Locals** and **Watch 1**.
- Click on the tab **Watch 1**.

- Click on the highlighted field under the label **Name**, type in the variable name f0, and press Enter. In the field under the label **Value**, you should see the number 1000, which is the frequency of the observed sinusoid.
- Click on the value 1000 and change it to 2000. You should see a 2 kHz sinusoid on the oscilloscope. Note that the processor was still running.
- Repeat above procedure for amplitude and increase the amplitude. Do not increase amplitude beyond 32767. The range of output values is limited from –32768 to + 32767 due to 16-bit codec. Do not attempt to send more than 16 bits to codec. The on-board codec uses a 2's complement format. Verify the increase in amplitude.

**Step 4: Applying the Slider Gel File**

The General Extension Language (gel) is an interpretive language similar to C. It allows you to change a variable such as amplitude, sliding through different values while the processor is still running. All variables must first be defined in your program.

- Select **File** > **Load gel** and open **amplitude.gel** that you created earlier.
- Select **gel** > **Sine Amplitude**. This should bring out a **slider window** with a minimum value of 10000 set for amplitude.
- Press **up-arrow** key to increase amplitude or use mouse to move the slider amplitude value. Verify the increase in amplitude on the oscilloscope.
- Two sliders can be readily used — one to increase amplitude and the other to change frequency.

### 7.2.6 Experimenting with the 'C6711 DSK as a Sine Wave Generator

**Exercise 4: This experiment tests the 'C6711 DSK as a sine wave generator using polling and consists of three steps given below. Please follow the instructions carefully to complete the experiment.**

This section has three purposes: to demonstrate how to reuse a previously created project, to create a real-time communication link between the CPU and codec using **polling**, and to generate a sinusoid using a **lookup table**. To create the project **sine lookup_poll**, follow these instructions:

**Step 1: Creating, Deleting, and Adding Files**

- Create a folder in Windows Explorer to store this project (e.g., create the folder **c:\ti\myprojects\sine_lookup_poll**).
- Copy the files **sine_gen.pjt** and **sine_gen.c**, from the previous project, into your newly created folder.

- Change the names of **sine_gen.pjt** and **sine_gen.c** to **sine_lookup_poll.pjt** and **sine_ lookup_poll.c**, respectively.
- Open Project **sine_lookup_poll.pjt** in CCS. When the window appears that says CCS cannot find the file **sine_gen.c**, select **Ignore**. Depending upon the drive where you created this new folder, you might have to include **rts6701. lib** file again.
- Delete **sine_gen.c** by selecting **sine_gen.c** in left window and pressing delete. Add the renamed C source code file **sine_lookup_poll.c** to the project by selecting **Project**.
- Select **Add Files to Project**, then select **sine_lookup_poll.c** and click **Open**. Also delete the **vectors_11.asm** file and add the other vectors file, **vectors.asm**, located in your **c:ti\tutorial\dsk6711\Hello1** folder.

**Step 2: Building and Running Files**

- In CCS, go to **Build** options. Click on the **Linker** tab and change the word **sine_gen** in **Debug\sine_gen.out** to **sine_lookup_poll.out** in the field output filename.
- In CCS, double click on **sine_lookup_poll.c** in the lefthand window. Change the C code to the following:

```
short sine_table[8] = {0,14142,20000,14142,0,-14142,
-20000,-14142};
short loop ;
short amplitude = 1;
void main()
{
loop=0;
comm_poll();
while(1)
{
output_sample(sine_table[loop]);
if (loop < 7) ++loop;
else loop = 0;
}
}
```

- Add comments to your code where appropriate and save the file in CCS.
- Now, build your project by clicking on the **rebuild all** button (the button with three red arrows).

- Before loading a new program onto the DSK, it is best to reset the DSP. This can be done within CCS by selecting the **Debug** pull-down menu and then selecting **Reset CPU**.
- Once the DSP is reset, load your new program onto the DSK and observe a 1KHz sine wave on an oscilloscope with an amplitude of approximately 1.75 volts. Notice that the sine wave algorithm is now coded within the infinite while loop (while(1)).

This is the general structure for polling. In both polling- and interrupt-based programs, the algorithm must be small enough to execute within 0.125ms (at an 8 kHz rate) in order to maintain a constant output to the on-board codec. Algorithms can be coded under either scheme, using polling or interrupts.

**Step 3: Alternative to Computing Sine Values**

As an alternative to computing sine values, a lookup table for generating a sinusoid may be used. The advantage is that because the same values are being repeatedly sent to the on-board codec, they may be stored locally in memory, so they do not need to be constantly recalculated. An example of this would be storing the 256 twiddle factors of a 256-point FFT algorithm. For generating sinusoids of various frequencies, a large sine table (e.g., 1000 points or more) may be created. The frequency of the sinusoid can be changed by incrementing the counter variable **loop** by any integer smaller than the length of the table at each interval. In the previous code, the command **++loop**; incremented the counter by one, which in C is equivalent to writing either **loop += 1**; or **loop = loop + 1**;. Also, in MATLAB, the values of sine_table[8] were generated by the command 20000*$sin(2\pi\ 0{:}7/N)$, where in this case $N = 8$. Since the number of points is small, these 8 values were included directly into the C source code. For larger sine tables, it is recommended that you store the values in a header file (**extension.h**) and include the file in the beginning part of your program. The number of points determines the frequency of the sampled sinusoid, 8000/N Hz. In this example, the number of points, $N$, was 8, so the sinusoid frequency was $8000/N$ = 1kHz. As a final note, the counter variable, **loop**, needs to be reset to zero only after it increments past $N - 1$. In the previous program, this value was 7, since there were 8 samples of a sinusoid.

### 7.2.7 Experimenting with the 'C6711 DSK for Math Operations

**Exercise 5: This experiment tests the 'C6711 DSK as a math calculator to obtain the dot product of two arrays, and the procedure is given below in a series of three steps. Please follow the instructions carefully in all the steps to complete the experiment.**

Operations such as addition, subtraction, and multiplication are the key operations in a digital signal processor. A very important application is the multiply/accumulate, which is useful in a number of applications requiring digital filtering, correlation, and spectrum analysis.

**Step 1: Creating the Header and C Code Files**

- Create a header file with the following listing and save it as **dotp5.h**.

```
//dotp5.h Header file with two arrays of numbers
#define x_array 1, 2, 3, 4,1
#define y_array 0, 2, 4, 6, 1
```

- Create a C code file with the following listing and save it as **dotp5.c**

```
//dotp5.c Multiplies two arrays, each with 5 numbers
int dotp(short *a, short*b, int ncount); //function
prototype
# include <stdio.h>              //for printf
# include "dotp5.h"              //data file of numbers
# define count 5                 //# of data in each array
short x[count] = {x_array};      //declare 1st array
short y[count] = {y_array};      //declare 2nd array
main ()
{
int result = 0;                  //result sum of products
result = dotp(x,y,count);        //call dotp function
printf("result =% d (decimal) \ n," result); //print
                                 result

}
int dotp(short *a, short*b, int ncount) //dot product
function
int sum = 0;                     //init sum
int i;
for (i= 0; i<ncount ; i++)
sum += a[i] * b[i];              //sum of products
return (sum);                    //return sum as result
}
```

The C source file **dotp5.c** takes the sum of product of two arrays, each with five numbers, contained in the header file **dotp5.h**. The support functions for interrupts are not needed here. The vector file used is less extensive.

**Step 2**

- Create and build this project as **dotp5.pjt** and add the following files to the project:
  - **dotp5.c** : source file
  - **vectors.asm** : vector file defining entry address C_int00
  - **C6xdsk.cmd** : linker file
  - **rts6701.lib** : file

Do not add any include file using **Add files to project**, because they are added by selecting **Project > scan all dependencies**. The header file **stdio.h** is needed due to **printf** statement in program **dotp5.c** to print the result. The header file **dotp5.h** is included upon scanning all dependencies.

**Step 3: Running the Program**

- Load and run the program. Verify the result of the dot product.

## 7.3 End Notes

The first lab was used to learn how to create a project and implement it on the DSK. In all real-time DSP algorithm implementations, the processing rate of a digital signal processing system is very important. For this lab, only an 8 KHz rate was used to implement algorithms. For more introductory information about the 'C6711 see references[3, 4] and Appendix E of this book.

## References

1. DSP Chips — Internet Resources, http://www.eg3.com. Mentor Graphics, 2004.
2. Texas Instruments homepage, http://www.ti.com. 2004.
3. Tretter, S.A., *Communication Design Using DSP Algorithms: With Laboratory Experiments for the TMS320C6701 and TMS320C6711*, Kluwer Academic/Plenum Publishers, New York, 2003.
4. Chassaing, R., *DSP Applications Using C and the TMS320C6x DSK*, Wiley, New York, 2002.
5. Chassaing, R., *DSP Applications Using C and the TMS320C31 DSK*, Wiley, New York, 1999.
6. Kehtarnavaz, N. and Keramat, M., *DSP System Design Using the TMS320C6000*, Prentice Hall, Upper Saddle River, NJ, 2001.
7. Texas Instruments, *TMS320C6000 Programmer's Guide, SPRU198D*, Dallas, TX, 2000.

8. Texas Instruments, *CPU and Instruction Set Reference Guide, SPRU189F,* Dallas, TX, 2000.
9. Texas Instruments, *TMS320C6000 Code Composer Studio User's Guide,* Dallas, TX, *SPRU328B*, 2001.
10. Texas Instruments, *TMS320C6000 Peripherals, SPRU190D*, Dallas, TX, 2001.

# 8

# *DSP Hardware Design II*

## 8.1 Overview of Practical DSP Applications in Communication Engineering

The TMS320C6711 DSP Starter Kit (or 'C6711 DSK) provides system design engineers with an easy-to-use, cost-effective way to take their high-performance TMS320C6000 designs from concept to production. As was extensively discussed in the previous chapter — and is also covered in Appendix E — the 'C6711 DSK is powerful enough to use for fast development of networking, communications, imaging, and other applications.

The 'C6711 DSK has the capability of real-time signal processing operations, the most important of which is *digital filtering*. Filtering is one of the most widely used applications in communications engineering.[1] Some of the practical applications of filtering are listed below.

- **Demodulation of AM and FM signals**: Low-pass filtering is used to recover baseband audio or video signal from the modulated signal.
- **Stereo generation**: In stereo systems, the basic audio signal is separated into low frequency and high frequency components using filter banks, amplified, and then synthesized to generate the stereo signal.
- **Filtering of noise**: Communications signals such as audio and video signals are corrupted by various sources of noise during propagation through communication channels. Filters are very useful in signal restoration and signal enhancement.

In this laboratory, students will design, simulate, and implement three important filtering applications using the 'C6711 DSK. The laboratory will cover a wide spectrum of software and hardware tools, including using MATLAB to design filters, programming the 'C6711 DSK to implement the filters, and finally, using signal sources and measuring equipment to test the overall applications.

## 8.2 Filtering Application to Extract Sinusoidal Signal from a Combination of Two Sinusoidal Signals

**Exercise 1: This two-signal filtering application consists of a series of six steps. Please follow the instructions carefully to successfully complete the experiment.**

In many communications applications, including wireless and cellular, it is often required to separate *two baseband signals*, $s_1(t)$ and $s_2(t)$, with different frequencies, $f_1$ Hz and $f_2$ Hz, respectively. The essential filtering process is shown in Figure 8.1, where a *bandstop filter* can be used to filter out the signal $s_1(t)$ or $s_2(t)$.

The procedure to implement the system, shown in Figure 8.1, is divided into six experimental steps, with each step being very important to the overall application.

**Step 1: Design of digital bandstop filter using MATLAB**

There are several ways to design digital filters using MATLAB, which were discussed in Chapter 5. However, we will focus on the method that is available even on the simplest MATLAB student version. Real-time digital filters can be implemented using the following protocol:

- Given a desired analog frequency response, $H_d(j\Omega)$, convert the latter response to the corresponding digital frequency response, $H_d(e^{j\omega})$, using the transformation $\omega = \Omega T$, where $T$ (sec.) is the sampling interval. Note that $T = 1/f_s$, where $f_s$ is the sampling frequency (Hz). The default sampling rate in the DSK is 8 KHz.

In this application, we have to design a bandstop filter with a center frequency of $f_1$ Hz, and a bandwidth of $\Delta f$ Hz. Hence, the cutoff frequencies

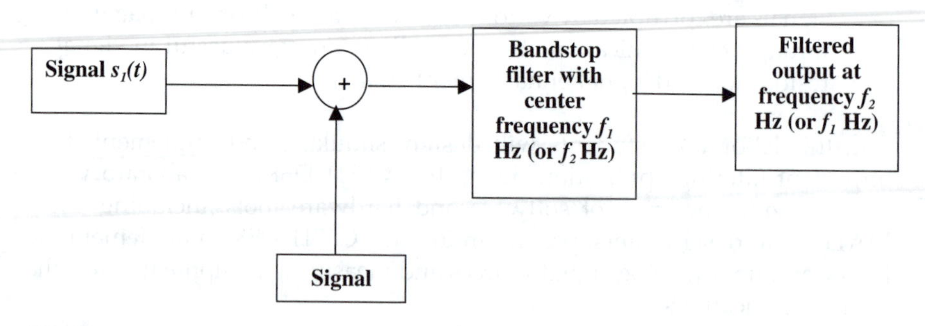


**FIGURE 8.1**
Signal filtering from a combination of two signals of different frequency.

of the desired bandstop filter are $f_l = f_1 - \Delta f/2$ (lower cutoff frequency) and $f_u = f_1 + \Delta f/2$ (upper cutoff frequency).

- Obtain the desired *Nth* order FIR digital filter coefficients $h(n)$, $0 \le n \le N$ using MATLAB. The various MATLAB commands for FIR digital filter design are given in Section 5.3.2 of this book. Examples of these commands are **fir1**, **fir2**, and **Remez**, in which the appropriate windowing function should also be specified. A sample program using the **fir1** command is given below.

```
% MATLAB Program to calculate the FIR bandstop filter
coefficients
N = 50; specifies the filter order (50)
fs = 8000; specifies the sampling frequency (8 KHz)
f = [f1 fh]; specifies the analog filter cutoff frequency
vector in Hz.
wn = 2*pi*f/fs; transforms the analog cutoff frequency
vector, f Hz, to digital cutoff frequency vector, wn, rad.
wn = wn/pi; normalizes the digital cutoff frequency vector
(MATLAB) requirement)
h = fir1(N, wn, 'stop'); calculates the bandstop FIR
filter coefficients
```

*Note:* If no window function is specified, as in the program above, then MATLAB uses the Hamming window.

Once the required filter coefficients, $h(n)$, $0 \le n \le N$, are obtained, a coefficient file, **bandstop.cof** is created as shown below.

```
/*bandstop.cof FIR bandstop filter coefficients file*/

#define N 51 /*length of filter*/

short hbs[N]=
{ h(0),h(1),..........................h(10),
  h(11),h(12),..........................h(20),
  h(21), h(22), ..........................h(30),
  h(31), h(32), ..........................h(40),
  h(41), h(32), ..........................h(50)
};
```

**Step 2: Create C program to implement bandstop filter on the '6711 DSK**

- The C program basically executes the filter operation, defined by the following convolution equation, which was discussed initially in Chapter 2.

$$y(n) = \sum_{k=-\infty}^{\infty} x(k)h(n-k) \tag{8.1}$$

The C language filter program **fir.c** is given below. Some of the important features of the program are as follows:

- The program **fir.c** is a very generic program and can be used for the implementation of *any kind* of FIR filter, as defined by the operation in Equation 8.1. It is only the coefficient file, **bandstop.cof**, which has to be changed according to the type of filter. Ultimately, it is only the numbers within the coefficient file that govern the nature of the filter, which is one of the remarkable advantages in the implementation of digital systems.
- Hence, the same program, **fir.c**, can be used for the other two applications in this laboratory, taking care to include the appropriate coefficient file for the application.

---

```
//fir.c FIR filter. Include coefficient file with length N

#include "bandstop.cof"        //coefficient file
int yn = 0;                    //initialize filter's output
short dly[N];                  //delay samples

interrupt void c_int11()       //ISR
{
   short i;

   dly[0] = input_sample(); //new input @ beginning of
                              buffer
   yn = 0;                  //initialize filter's output
   for (i = 0; i< N; i++)
     yn += (h[i] * dly[i]); //y(n) += h(i)* x(n-i)
   for (i = N-1; i > 0; i–) //starting @ end of buffer
   dly[i] = dly[i-1];       //update delays with data
                              move
```

```
   output_sample(yn >> 15);//scale output filter
   return;
}

void main()
 {
   comm_intr();                //init DSK, codec, McBSP
   while(1);                   //infinite loop
 }
```

**Step 3: Setting up the 6711 DSK for filter implementation**

The steps for this exercise are given in detail in Section 7.2. However, the main instructions are again summarized below.

- **Initial setting up of the equipment**
  - Connect the parallel printer port cable between the parallel port on the DSK board (J2) and the parallel printer port on the back of the computer.
  - Connect the 5V power supply to the power connector next to the parallel port on the DSK board (J4). You should see 3 LEDs blink next to some dip switches.

  Once the DSK board is connected to your PC and the power supply has been connected, you can start CCS.

  - To do this, click on **Start**, go to **Program**, and then go to **Texas Instruments**, then **Code Composer Studio DSK Tools 2 ('C6000)**, select **CCStudio**.
  - Or from Desktop click on **CCS-DSK 2 ('C6000)**.
- **Creating a new project file**
  - In CCS, select **Project** and then **New**. A window named **Project Creation** will appear.
  - In the field labeled **Project Name**, enter **filtering_twosignals**. In the field **Location**, click on the right side of the field and browse to the folder **c:\ti\myprojects\ filtering_twosignals**.
  - In the field **Project Type**, verify that **Executable (.out)** is selected, and in the field **Target**, verify that **TMS320C67XX** is selected.
  - Finally, click on **Finish**. CCS has now created a project file **filtering_twosignals.pjt**, which will be used to build an executable program. This file is stored in the folder **c:\ti\myprojects\ filtering_twosignals.** The**.pjt** file stores project information on build options, source filenames, and dependencies.

- **Loading the support files**

  Add the following support files to the project **c:\ti\myprojects\filtering_twosignals**. Details on the functions of these support files are given in Section 7.2. Note that you need to scan all dependencies after all files have been added, including source files, in order to include the header files. Header files cannot be added to a project.

  - **C6xdsk.cmd**
  - **C6x.h**
  - **C6xdskinit.c**
  - **Vectors_11.asm**
  - **C6xdsk.h**
  - **C6xinterrupts.h**
  - **C6xdskinit.h**
  - **rts6701.lib**

- **Loading the program files**

  Add the C source files **fir.c** and the filter coefficient file **bandstop.cof** to the project **c:\ti\myprojects\filtering_twosignals**.

**Step 4: Hardware setup for the filtering of two sinusoidal signals**

- Generate a mixed signal (using a BNC TEE junction) consisting of two sinusoidal signals of frequency 1.5 KHz and 3 KHz, both with amplitude of 0.5 volts, as shown in Figure 8.2. (Note that this step will require two HP 3324A signal generators.) Verify the mixed output signal, both in time and frequency domains, using the HP 35665A Dynamic Signal Analyzer.
- Connect the mixed signal output to the *input* of the 'C6711 DSK, and connect the output of the 'C6711 DSK to **Channel 1** of the HP 35665A Dynamic Signal Analyzer, as shown in Figure 8.2.

The experimental setup is complete for measurements.

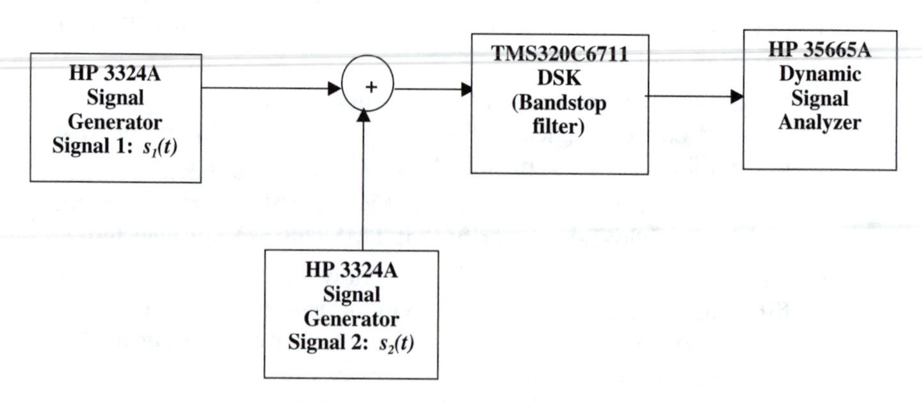


**FIGURE 8.2**
Experimental setup for signal filtering of a combination of two signals.

**Step 5: Running the DSK and making measurements**

- Follow the steps shown in steps 1–3, to implement a digital bandstop filter on the 'C6711 DSK. The bandstop filter should have a center frequency of 1.5 kHz and bandwidth of 100 Hz.
- Go to **Project** pull-down menu in the **CCS window**, and then select **Build** (or press the button with three red down arrows on the top toolbar in the CCS window). A new subwindow will appear on the bottom of the CCS window. When building is complete, you should see the following message in the new subwindow:

  ```
  Build Complete,
  0 Errors, 0 Warnings, 0 Remarks
  ```

  The following executable file will be created:

  ```
  c:\ti\myprojects\filtering_twosignals\Debug
  \filtering_twosignals.out
  ```

- Click on the **Debug** pull-down menu and select **Reset CPU**.
- Then, to load the program onto the DSK, click on the **File** pull-down menu and select **Load Program**.
- In the new window that appears, double click on the folder **Debug**, select the **filtering_twosignals.out** file, and click on **Open**.
- In CCS, select the **Debug** pull-down menu and then select **Run**, or simply click on the top **running man** on the left side toolbar. You should now see the filtered output with a predominant peak at 3 kHz on the Signal Analyzer. However, there may be a small component at 1.5 kHz, *hence measure the power level (dBm) at both 1.5 kHz and 3 kHz.*

**Step 6: Design of DSK to extract signal with frequency of 2 KHz**

Repeat the procedure in the step 2 and implement a bandstop filter centered at 3 kHz and a bandwidth of 0.4 kHz. Observe the filtered output on the HP 35665A Dynamic Signal Analyzer, and check that there is a significant peak at 1.5 KHz. However, measure the power level (dBm) at both 1.5 kHz and 3 kHz.

## 8.3 Filtering Application to Extract Sinusoidal Signal from a Noisy Signal

**Exercise 2: This noisy signal filtering application consists of a series of five steps. Follow the instructions carefully to complete the experiment.**

All communications systems face the common problem of noise, in greater or lesser measure. As shown in Figure 8.3, the simplest form of noise is

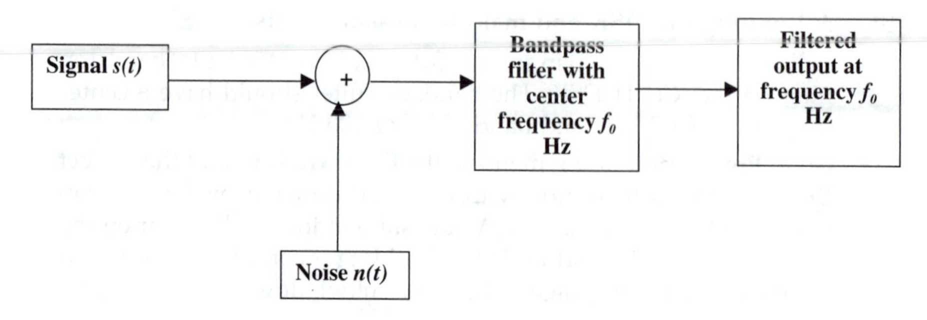


**FIGURE 8.3**
Signal filtering from a combination of signal and noise.

additive noise $n(t)$, which adds on to the transmitted signal $s(t)$ at a frequency of $f_0$ Hz. Several methods have been developed to tackle the problem of noise removal from the corrupted signal $y(t) = s(t) + n(t)$. The commonly used methods include autocorrelation, and filtering.

**Step 1: Design of bandpass filter using MATLAB**

The filter design procedure is very similar to the bandstop filter design in the previous application in Section 8.2. However, the steps are retraced for convenience.

- Given the desired analog frequency response, $H_d(j\Omega)$, convert to the corresponding digital frequency $H_d(e^{j\omega})$, using the transformation $\omega = \Omega T$, where $T$ (sec.) is the sampling interval. Note that $T = 1/f_s$, where $f_s$ is the sampling frequency (Hz). The default sampling rate in the DSK is 8 KHz.

  In this application, we have to design a bandpass filter with a center frequency of $f_1$ Hz and a bandwidth of $\Delta f$ Hz. Hence, the cutoff frequencies of the desired bandpass filter are $f_l = f_1 - \Delta f/2$ (lower cutoff frequency) and $f_u = f_1 + \Delta f/2$ (upper cutoff frequency).
- Obtain the desired *Nth* order FIR digital filter coefficients $h(n)$, $0 \leq n \leq N$ using MATLAB. The various MATLAB commands for FIR digital filter design are given in Section 5.3.2 of this book. Examples of these commands are **fir1**, **fir2**, and **Remez**, in which the appropriate windowing function should also be specified. A sample program using the **fir1** command is given below:

```
% MATLAB Program to calculate the FIR bandpass filter
coefficients
N = 50            ; specifies the filter order (50)
fs = 8000         ; specifies the sampling frequency (8 KHz)
f =[f1 fh]        ; specifies the analog filter cutoff
                    frequency vector in Hz.
```

```
wn = 2*pi*f/fs ; transforms the analog cutoff frequency
                 vector, f Hz, to digital cutoff
                 frequency vector, wn, rad.
wn = wn/pi     ; normalizes the digital cutoff frequency
                 vector
h = fir1(N, wn); calculates the FIR filter coefficients
```

Once the required filter coefficients, $h(n)$, $0 \le n \le N$, are obtained, a coefficient file, **bandpass.cof** should be created as shown below:

```
/*bandpass.cof FIR bandpass filter coefficients file*/

#define N 51 /*length of filter*/

short hbp[N]=
{ h(0),h(1),h(2),h(3),h(10),
  h(11),h(12),h(2),h(3),h(20),
  h(21), h(22),h(2),h(3),h(30),
  h(31), h(32),h(2),h(3),h(40),
  h(41), h(32),h(2),h(3),h(50)
};
```

**Step 2: C program to implement bandpass filter on the 'C6711 DSK**

The same generic filter program **fir.c**, which was explained in Section 8.2, step 2, can be utilized for this application. The program listing is given below, however, taking care to include the appropriate *bandpass* coefficient file for the application.

```
//Fir.c FIR filter. Include coefficient file with length N

#include "bandpass.cof"        //coefficient file
int yn = 0;                    //initialize filter's output
short dly[N];                  //delay samples
interrupt void c_int11()       //ISR
{
   short i;

   dly[0] = input_sample();  //new input @ beginning of
                               buffer
   yn = 0;                     //initialize filter's output
```

```
    for (i = 0; i< N; i++)
      yn += (h[i] * dly[i]);  //y(n) += h(i)* x(n-i)
    for (i = N-1; i > 0; i—)  //starting @ end of buffer
      dly[i] = dly[i-1];      //update delays with data
                                move

    output_sample(yn >> 15);  //scale output filter
    return;
}
void main()
 {
    comm_intr();              //init DSK, codec, McBSP
    while(1);                 //infinite loop
 }
```

**Step 3: Setting up the 6711 DSK for filter implementation**

The steps for this exercise are given in detail in Section 8.2. However, the main instructions are again summarized below.

- **Initial setting up of the equipment**
  - Connect the parallel printer port cable between the parallel port on the DSK board (J2) and the parallel printer port on the back of the computer.
  - Connect the 5V power supply to the power connector next to the parallel port on the DSK board (J4). You should see 3 LEDs blink next to some dip switches.

  Once the DSK board is connected to your PC and the power supply has been connected, you can start CCS.

  - To do this, click on **Start**, go to **Program**, and then go to **Texas Instruments**, then **Code Composer Studio DSK Tools 2 ('C6000)**, select **CCStudio**.
  - Or from Desktop click on **CCS-DSK 2 ('C6000)**.
- **Creating a new project file**
  - In CCS, select **Project** and then **New**. A window named **Project Creation** will appear.
  - In the field labeled **Project Name**, enter **filtering_signal&noise**. In the field **Location**, click on the on the right side of the field and browse to the folder **c:\ti\myprojects\filtering_signal&noise**.
  - In the field **Project Type**, verify that **Executable (.out)** is selected, and in the field **Target**, verify that **TMS320C67XX** is selected.

  - Finally, click on **Finish**. CCS has now created a project file **filtering_signal&noise.pjt**, which will be used to build an executable program. This file is stored in the folder **c:\ti\myprojects\filtering_signal&noise.** The.**pjt** file stores project information on build options, source filenames, and dependencies.
- **Loading the support files**

  Add the following support files to the project **c:\ti\myprojects\filtering_signal&noise**. Remember header files are included by scanning all dependencies, after all files have been added, including source files.
  - **C6xdsk.cmd**
  - **C6x.h**
  - **C6xdskinit.c**
  - **Vectors_11.asm**
  - **C6xdsk.h**
  - **C6xinterrupts.h**
  - **C6xdskinit.h**
  - **rts6701.lib**
- **Loading the program files**

  Add the C source files **fir.c** and the filter coefficient file **bandpass.cof** into the project directory **c:\ti\myprojects\filtering_signal&noise**.

**Step 4: Hardware setup for the filtering of two sinusoidal signals**

- Connect the experimental setup as shown in Figure 8.4
- Generate a sinusoidal signal, *s*(*t*), of amplitude 0.5 volts and frequency of 3 KHz, using the HP 3324A Signal Generator. Check the output of the generator on the HP 35665A Dynamic Signal Analyzer, and make a plot of the pure sinusoidal signal on the printer.

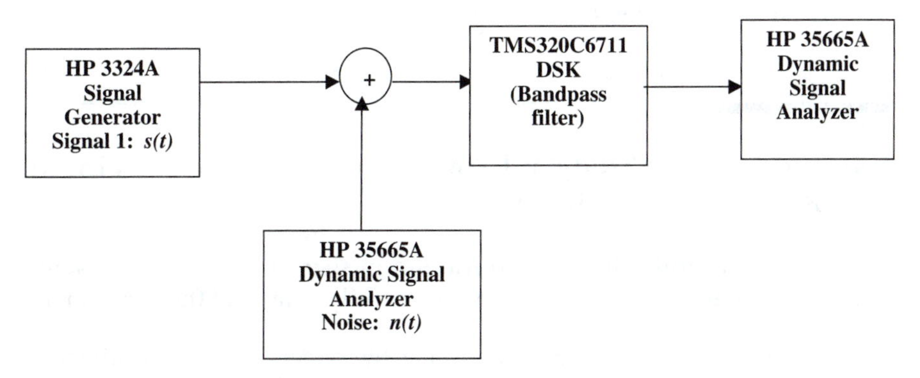


**FIGURE 8.4**
Experimental setup for filtering of a noisy signal.

- Generate a random noise signal, n(t), using the HP 35665A Dynamic Signal Analyzer. Set the noise level at 0.05 volt rms.
- Combine the signal s(t) and the noise n(t), as shown in Figure 8.4, and check the noisy output on the HP 35665A Dynamic Signal Analyzer. Plot the noisy output on the printer.

The experimental setup is complete for measurements.

**Step 5: Running the DSK and making measurements**

- Follows the steps shown in steps 1–3, to implement a bandpass filter on the 'C6711 DSK. The bandpass filter should have a center frequency of 3 kHz and bandwidth of 100 Hz.
- Go to **Project** pull-down menu in the **CCS window**, and then select **Build** (or press the button with three red down arrows on the top toolbar in the CCS window). A new subwindow will appear on the bottom of the CCS window. When building is complete, you should see the following message in the new subwindow:

  ```
  Build Complete,
  0 Errors, 0 Warnings, 0 Remarks
  ```

  The executable file **c:\ti\myprojects\filtering_signal&noise\Debug\filtering_signal&noise.out** will be created.
- Click on the **Debug** pull-down menu and select **Reset CPU**.
- Then, to load the program onto the DSK, click on the **File** pull-down menu and select **Load Program**.
- In the new window that appears, double click on the folder **Debug**, select the **filtering_signal&noise.out** file, and click on **Open**.
- In CCS, select the **Debug** pull-down menu and then select **Run**, or just simply click on the top **running man** on the left side toolbar. You should now see the filtered output on the Signal Analyzer.
- Verify the filtered output on HP 35665A Dynamic Signal Analyzer. Make a plot of the filtered output on the printer.

## 8.4 Comparative Study of Using Different Filters on an Input Radio Receiver Signal

**Exercise 3: This multifiltering application consists of a series of five steps. Follow the instructions carefully to successfully complete the experiment.**

In this laboratory, we will study and hear the effects of different filters on voice and music signals coming from a common AM/FM radio receiver. The experimental setup is shown in Figure 8.5. The 'C6711 DSK will again be

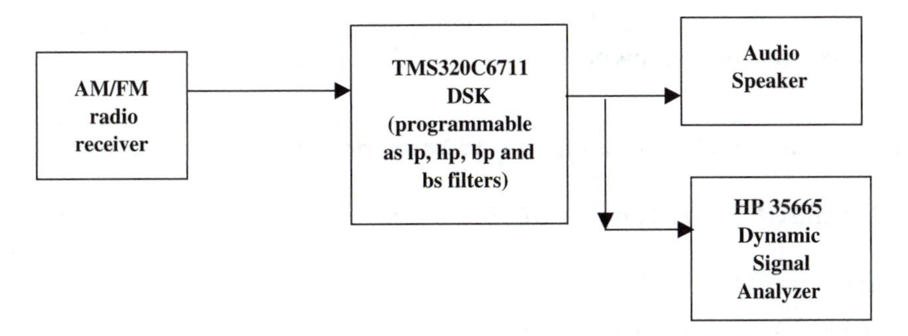


**FIGURE 8.5**
Experimental setup for study of filtering effects on speech and music signals.

used to implement the different types of filtering (low-pass, high-pass, band-pass, and bandstop). However, one new concept will be the use of the **.gel** file, or **slider** file, which will enable us to step through the four different types of FIR filters. However, before the actual hardware testing, there are important design steps, which are explained below.

### Step 1: Design of four filter types using MATLAB

The filter design procedure for this experiment is very similar to the previous two experiments, described in Section 8.2 and Section 8.3 However, we need to simultaneously design four kinds of filters, using the following MATLAB commands, and store the coefficients in their respective ***.cof** files.

- **Low-pass filter**
  Design a digital *low-pass* filter, having a cutoff frequency of 1.5 kHz, using the program below.

```
% MATLAB Program to Calculate the FIR Low-Pass Filter
Coefficients
N = 50          ; specifies the filter order (50)
fs = 8000       ; specifies the sampling frequency (8 KHz)
f= 1500         ; specifies the analog filter cutoff
                  frequency in Hz.
wn = 2*pi*f/fs ; transforms the analog cutoff frequency,
                  f Hz, to digital cutoff frequency, wn,
                  rad.
wn = wn/pi      ; normalizes the digital cutoff frequency
                  vector
h = fir1(N,wn) ; calculates the FIR low-pass filter
                  coefficients
```

Once the required filter coefficients, $h(n)$, $0 \leq n \leq N$, are obtained, a coefficient file, **lowpass.cof** is created as shown below.

```
/*lowpass.cof FIR lowpass filter coefficients file*/

#define N 51 /*length of filter*/

short hlp[N]=
{ h(0),h(1),..........................h(10),
  h(11),h(12),..........................h(20),
  h(21), h(22), ..........................h(30),
  h(31), h(32), ..........................h(40),
  h(41), h(32), ..........................h(50)
};
```

- **High pass filter**

  Design a digital *high-pass* filter, having a cutoff frequency of 1.5 kHz, using the program below.

```
% MATLAB Program to Calculate the FIR High-Pass Filter
Coefficients
N = 50                 ; specifies the filter order (50)
fs = 8000              ; specifies the sampling frequency
                         (8 KHz)
f = 1500               ; specifies the analog filter
                         cutoff frequency in Hz.
wn=2*pi*f/fs           ; transforms the analog cutoff
                         frequency, f Hz, to
                       ; digital cutoff frequency, wn,
                         rad.
wn=wn/pi               ; normalizes the digital cutoff
                         frequency vector
h = fir1(N,wn,'high')  ; calculates the FIR high-pass
                         filter coefficients
```

Once the required filter coefficients, $h(n)$, $0 \leq n \leq N$, are obtained, a coefficient file, **high pass.cof** is created as shown below:

```
/*highpass.cof FIR highpass filter coefficients file*/

#define N 51 /*length of filter*/

short hhp[N]=
{ h(0),h(1),..........................h(10),
  h(11),h(12),..........................h(20),
  h(21), h(22), ..........................h(30),
  h(31), h(32), ..........................h(40),
  h(41), h(32), ..........................h(50)
};
```

- **Bandpass filter**

  Design a digital *bandpass* filter, having a *center* frequency of 1.5 kHz and bandwidth of 200 Hz, using the program below. Because the bandwidth of the filter is 200 Hz, the cutoff frequencies of the band-pass filter are 1.4 kHz and 1.6 kHz, respectively.

```
% Program to Calculate the FIR Bandpass Filter
Coefficients
N = 50             ; specifies the filter order (50)
fs = 8000          ; specifies the sampling frequency (8 KHz)
f =[1400 1600]     ; specifies the analog filter cutoff
                     frequency vector in Hz.
wn = 2*pi*f/fs     ; transforms the analog cutoff frequency
                     vector, f Hz, to ;digital cutoff
                     frequency vector, wn, rad.
wn = wn/pi         ; normalizes the digital cutoff frequency
                     vector
h = fir1(N,wn)     ; calculates the FIR bandpass filter
                     coefficients
```

Once the required filter coefficients, $h(n)$, $0 \le n \le N$, are obtained, a coefficient file, **bandpass.cof** is created as shown below.

```
/*bandpass.cof FIR bandpass filter coefficients file*/

#define N 51 /*length of filter*/

short hbp[N]=
{ h(0),h(1),..........................h(10),
```

```
  h(11),h(12),..........................h(20),
  h(21), h(22), ..........................h(30),
  h(31), h(32), ..........................h(40),
  h(41), h(32), ..........................h(50)
};
```

- **Bandstop filter**

  Design a digital *bandstop* filter, having a *center* frequency of 1.5 kHz and bandwidth of 200 Hz, using the program below. Because the bandwidth of the filter is 200 Hz, the cutoff frequencies of the bandstop filter are 1.4 kHz and 1.6 kHz, respectively.

```
% Program to Calculate the FIR Bandstop Filter
Coefficients
N = 50                    ; specifies the filter order (50)
fs = 8000                 ; specifies the sampling frequency
                            (8 KHz)
f =[1400 1600]            ; specifies the analog filter
                            cutoff frequency vector in Hz.
wn = 2*pi*f/fs            ; transforms the analog cutoff
                            frequency vector, f Hz, to
                          ; digital cutoff frequency vector,
                            wn, rad.
wn = wn/pi                ; normalizes the digital cutoff
                            frequency vector
h = fir1(N,wn,'stop')     ; calculates the Fir bandstop
                            filter coefficients
```

Once the required filter coefficients, $h(n)$, $0 \leq n \leq N$, are obtained, a coefficient file, **bandstop.cof** is created as shown below.

```
/*bandstop.cof FIR bandstop filter coefficients file*/

#define N 51 /*length of filter*/

short hbs[N]=
{ h(0),h(1),..........................h(10),
  h(11),h(12),..........................h(20),
  h(21), h(22), ..........................h(30),
```

```
  h(31), h(32), ..........................h(40),
  h(41), h(32), ..........................h(50)
};
```

**Step 2: C program to implement four filters on the 6711 DSK**

In order to implement four different kinds of filters, we use a new program, **Fir4types.c**. The program listing is given below.

```
//Fir4types.c Four FIR Filters: Low-pass, High-pass,
Bandpass, Bandstop

#include "lowpass.cof"       //coeff file LP @ 1500 Hz
#include "highpass.cof"      //coeff file HP @ 1500 Hz
#include "bandpass.cof"      //coeff file BP @ 1500 Hz
#include "bandstop.cof"      //coeff file BS @ 1500 Hz
short FIR_number = 1;        //start with 1st LP filter
int   yn = 0;                //initialize filter's output
short dly[N];                //delay samples
short h[4][N];               //filter characteristics 3xN

interrupt void c_int11()     //ISR
 {
   short i;
   dly[0] = input_sample(); //newest input @ top of
                              buffer
   yn = 0;                   //initialize filter output
   for (i = 0; i< N; i++)
     yn +=(h[FIR_number][i]*dly[i]);//y(n) += h(LP#,i)*
                             //x(n-i)
   for (i = N-1; i > 0; i—) //starting @ bottom of buffer
     dly[i] = dly[i-1];      //update delays with data
                              move
   output_sample(yn >> 15); //output filter
   return;                   //return from interrupt
 }
void main()
{
   short i;
```

```
  for (i=0; i<N; i++)
  {
      dly[i] = 0;            //init buffer
      h[1][i] = hlp[i];      //start addr of lowpass1500
                               coeff
      h[2][i] = hhp[i];      //start addr of highpass1500
                               coeff
      h[3][i] = hbp[i];      //start addr of bandpass1500
                               coeff
      h[4][i] = hbs[i];      //start addr of bandstop1500
                               coeff
  }
  comm_intr();               //init DSK, codec, McBSP
  while(1);                  //infinite loop
}
```

In addition to the main FIR filter implementation program, **Fir4types.c,** given above, we also require a **gel** or **slider file,** which will enable us to step through the four different kinds of filters when applied to the input voice or music signal. The **FIR4types.gel** file is given below.

```
/*FIR4types.gel Gel file for 4 different filters:
LP,HP,BP,BS*/
menuitem "Filter Characteristics"
slider Filter(1,4,1,1,filterparameter)//*from 1 to 4,incr
                                         by 1*/
{
    FIR_number = filterparameter; //*for 4 FIR filters*/
}
```

**Step 3: Setting up the 6711 DSK for filter implementation**

The procedures for this step are identical to the ones given in the previous two experiments, described in Section 8.2 and Section 8.3.

- **Creating a new project file**
  - In CCS, select **Project** and then **New**. A window named **Project Creation** will appear.
  - In the field labeled **Project Name**, enter **filtering_audio**. In the field 'Location', click on the on the right side of the field and browse to the folder **c:\ti\myprojects\filtering_audio**.

- In the field **Project Type**, verify that **Executable (.out)** is selected, and in the field **Target**, verify that **TMS320C67XX** is selected.
- Finally, click on **Finish**. CCS has now created a project file **filtering_audio.pjt**, which will be used to build an executable program. This file is stored in the folder **c:\ti\myprojects\filtering_audio.** The **.pjt** file stores project information on build options, source filenames, and dependencies.

- **Loading the support files**

  Add the following support files to the project **c:\ti\myprojects\filtering_audio**. After adding all files, including source files, scan all dependencies to include header files.

  - **C6xdsk.cmd**
  - **C6x.h**
  - **C6xdskinit.c**
  - **Vectors_11.asm**
  - **C6xdsk.h**
  - **C6xinterrupts.h**
  - **C6xdskinit.h**
  - **rts6701.lib**

- **Loading the program files**

  Add the C source file **Fir4types.c** into the project. The four filter coefficient files **lowpass.cof, highpass.cof, bandpass.cof,** and **band-stop.cof** will be automatically included into the project upon scanning all dependencies.

**Step 4: Hardware setup for the filtering of audio signals**

- Connect the hardware components, as shown in Figure 8.5.
- Set the AM/FM radio receiver at a clearly received audio station, such as FM 92.5 or AM 1320, for example. It would be typical to try this experiment for one *voice* signal and one *music* signal.
- Connect the radio received **audio output** directly to the speaker and check the audio output on the speaker.
- Also take a printout of the frequency spectrum of the signal on the Signal Analyzer, in a frequency range of 0-25 KHz.
- Now connect the system again, as shown in Figure 8.5, to include the 'C6711 DSK.

**Step 5: Running the DSK and making measurements**

- Go to **Project** pull-down menu in the **CCS window**, and then select **Build** (or press the button with three red down arrows on the top toolbar in the CCS window). A new subwindow will appear on the bottom of the CCS window. When building is complete, you should see the following message in the new subwindow:

```
Build Complete,
0 Errors, 0 Warnings, 0 Remarks
```

The executable file **c:\ti\myprojects\filtering_audio\Debug\filtering_ audio.out** will be created.

- Click on the **Debug** pull-down menu and select **Reset CPU**.
- Then, to load the program onto the DSK, click on the **File** pull-down menu and select **Load Program**.
- In the new window that appears, double click on the folder **Debug**, select the **filtering_audio.out** file and click on **Open**.
- In CCS, select the **Debug** pull-down menu and then select **Run**, or simply click on the top **running man** on the left side toolbar. You should now see the filtered output on the Signal Analyzer.
- Load the **gel** file, **FIR4types.gel,** and verify the implementation of the four different FIR filters. In each filter case, listen to the audio output on the speaker and make a record of the changes heard in the incoming signal. Also, in each filter case, take a printout of the filtered audio output on the Signal Analyzer.

## References

1. Tretter, S.A., *Communication Design Using DSP Algorithms: With Laboratory Experiments for the TMS320C6701 and TMS320C6711*, Kluwer Academic/Plenum Publishers, New York, 2003.
2. Chassaing, R., *DSP Applications Using C and the TMS320C6x DSK*, Wiley, New York, 2002.
3. Kehtarnavaz, N. and Keramat, M., *DSP System Design Using the TMS320C6000*, Prentice Hall, Upper Saddle River, NJ, 2001.

# APPENDIX A

## *HP/Agilent 3324A/33250A Synthesized Function/Sweep Generators*

### A.1 Introduction

The HP 33250A,[1] which is the successor of the HP3324A, is a function, arbitrary waveform, and pulse generator in one instrument, with the most stable frequencies and lowest distortion of any function generator in its class. It also provides internal AM, FM, and FSK modulation capabilities, sweep and burst operation modes, and a color display. The 33250A provides easy access to standard sine, square, ramp, triangle, and pulse waveforms — plus custom waveforms can be created using the 200 MSa/s, 12-bit, 64 K-point arbitrary waveform function.

The variable-edge pulse function gives the user unmatched flexibility for design, verification, and test applications. The 33250A also includes GPIB and RS-232 interfaces standard and IntuiLink software to enable simple generation of custom waveforms. Some of the key features of this equipment are as follows:

- 80 MHz sine and square waveforms
- Ramp, triangle, pulse, noise, and DC waveforms
- 12-bit, 200 MSa/s, 64 K-point arbitrary waveforms
- AM, FM, and FSK modulation types
- Linear and logarithmic sweep and burst operation modes
- Graph mode for visual verification of signal settings
- GPIB and RS-232 interfaces included
- Built-in multiple-unit link for synchronous channels

Some material in this Appendix is reproduced with permission from Agilent Technologies Inc., Palo Alto, CA.

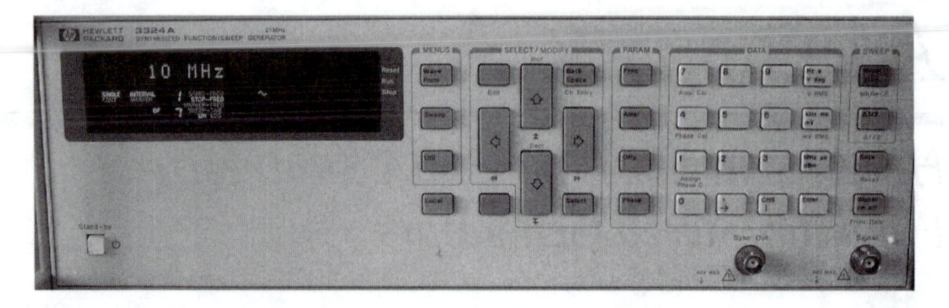


**FIGURE A.1**
HP3324A Synthesized Sweep and Function Generator. (Courtesy of Agilent Technologies Inc., Palo Alto, CA.)

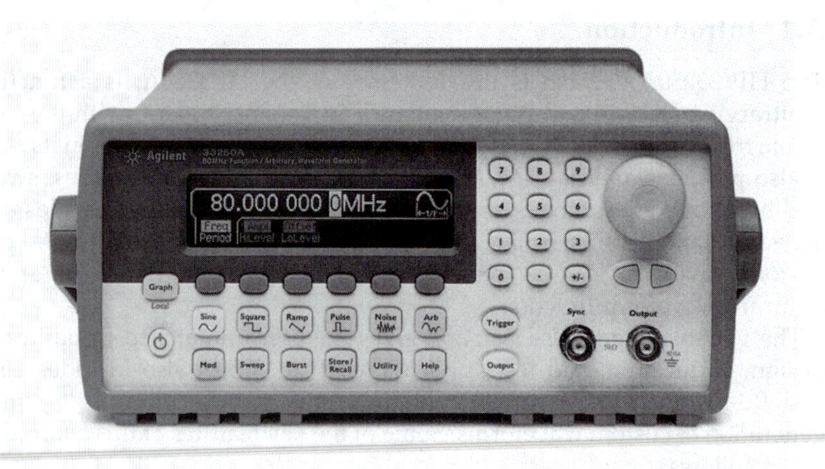


**FIGURE A.2**
HP33250A Synthesized Sweep and Function Generator. (Courtesy of Agilent Technologies Inc., Palo Alto, CA.)

The photograph of the earlier generation model HP3324A and the current model HP33250A are shown in Figure A.1 and Figure A.2.

## A.2 Technical Specifications of the Agilent HP3325A

### A.2.1 Waveforms

- Standard

  Sine, square, pulse, ramp, noise, sin(x)/x, exponential rise, exponential fall, cardiac, DC volts

- Arbitrary
  Waveform length: 1 to 64 K points
  Amplitude resolution: 12 bits (including sign)
  Repetition rate: 1 μHz to 25 MHz
  Sample rate: 200 MSa/s
  Filter bandwidth: 50 MHz
  Nonvolatile memory: Four 64 K waveforms

### A.2.2 Frequency Characteristics

Sine: 1 μHz to 80 MHz
Square: 1 μHz to 80 MHz
Pulse: 500 μHz to 50 MHz
Arbitrary: 1 μHz to 25 MHz
Ramp: 1 μHz to 1 MHz
White noise: 50 MHz bandwidth
Resolution: 1 μHz; except pulse, 5 digits
Accuracy (1 year): 2 ppm, 18°C to 28°C,
3 ppm, 0°C to 55°C

### A.2.3 Sinewave Spectral Purity

- Harmonic distortion

| | **≤3 Vpp1** | **≤3 Vpp** |
|---|---|---|
| DC to 1 MHz: | –60 dBc | 55 dBc |
| 1 to 5 MHz: | –57 dBc | –45 dBc |
| 5 to 80 MHz: | –37 dBc | –30 dBc |

- Total harmonic distortion
  DC to 20 kHz: <0.2% + 0.1 mVrms
- Spurious (nonharmonic)
  DC to 1 MHz: –60 dBc
  1 to 20 MHz: –50 dBc
  20 to 80 MHz: –50 dBc + 6 dBc/octave
- Phase noise (30 kHz band)
  10 MHz: <–65 dBc (typical)
  80 MHz : <–47 dBc (typical)

### A.2.4 Signal Characteristics

- Square wave
  Rise/fall time: <8 ns
  Overshoot: <5%
  Asymmetry: 1% of period + 1 ns
  Jitter (rms): <2 MHz 0.01% + 525 ps
  2 MHz 0.1% + 75 ps
  Duty cycle: 25 MHz 20.0% to 80.0%
  25 to 50 MHz 40.0% to 60.0%
  50 to 80 MHz 50.0% fixed
- Pulse
  Period: 20.00 ns to 2000.0 s
  Pulse width: 8.0 ns to 1999.9 s
  Variable edge time: 5.00 ns to 1.00 ms
  Overshoot: <5%
  Jitter (rms): 100 ppm + 50 ps
- Ramp
  Linearity: <0.1% of peak output
  Symmetry: 0.0% – 100.0%
- Arbitrary
  Minimum edge time: <10 ns
  Linearity: <0.1% of peak output
  Settling time: <50 ns to 0.5% of final value
  Jitter (rms): 30 ppm + 2.5 ns

### A.2.5 Output Characteristics

- Amplitude (into 50): 10 mV pp to 10 V pp
  Accuracy (at 1 kHz, >10 mV pp, Autorange): ±1% of setting ±1 mV pp
  Flatness (sinewave relative to 1 kHz, Autorange)
  <10 MHz: ±1% (0.1 dB)
  10 to 50 MHz: ±2% (0.2 dB)
  50 to 80 MHz: ±5% (0.4 dB)
  Units: Vpp, Vrms, dBm, high and low level
  Resolution: 0.1 mV or 4 digits
- Offset (into 50): ±5 Vpk AC + DC
  Accuracy: 1% of setting + 2 mV + 0.5% of amplitude

- Waveform output
  Impedance: 50 typical (fixed)
  >10 M (output disabled)
  Isolation: 42 Vpk maximum to earth
  Protection: short-circuit protected; overload automatically disables main output

### A.2.6 Modulation

- AM
  Carrier waveforms: sine, square, ramp, and arbitrary
  Modulation waveforms: sine, square, ramp, noise, and arbitrary
  Modulation frequency: 2 mHz to 20 kHz
  Depth: 0.0% to 120.0%
  Source: internal/external
- FM
  Carrier waveforms: sine, square, ramp, and arbitrary
  Modulation waveforms: sine, square, ramp, noise, and arbitrary
  Modulation frequency: 2 mHz to 20 kHz
  Deviation range: DC to 80 MHz
  Source: internal/external
- FSK
  Carrier waveforms: sine, square, ramp, and arbitrary
  Modulation waveform: 50% duty cycle square
  Internal rate: 2 mHz to 1 MHz
  Frequency range: 1 μHz to 80 MHz
  Source: internal/external
- External modulation input
  Voltage range: ±5 V full scale
  Input impedance: 10 k
  Frequency: DC to 20 kHz

### A.2.7 Burst

Waveforms: sine, square, ramp, pulse, arbitrary, and noise
Frequency: 1 μHz to 80 MHz3
Burst count: 1 to 1,000,000 cycles or infinite
Start/stop phase: –360.0° to +360.0°
Internal period: 1 ms to 500 s

Gate source: external trigger

Trigger source: single manual trigger, internal, external trigger

Trigger delay: N-cycle, infinite 0.0 ns to 85.000 sec

### A.2.8 Sweep

Waveforms: sine, square, ramp, and arbitrary

Type: linear and logarithmic

Direction: up or down

Start F/Stop F: 100 μHz to 80 MHz

Sweep time: 1 ms to 500 s

Trigger: single manual trigger, internal, external trigger

Marker: falling edge of sync signal (programmable)

### A.2.9 System Characteristics

- Configuration Times (typical)
  Function change
    Standard: 100 ms
    Pulse: 660 ms
    Built-in arbitrary: 220 ms
  Frequency change: 20 ms
  Amplitude change: 50 ms
  Offset change: 50 ms
  Select user arbitrary: <900 ms for <16K pts
  Modulation change: <200 ms
- Arbitrary Download Times GPIB/RS-232 (115Kbps)

| | Arb Length Binary | ASCII Integer | ASCII Real |
|---|---|---|---|
| 64K points | 48 sec | 112 sec | 186 sec |
| 16K points | 12 sec | 28 sec | 44 sec |
| 8K points | 6 sec | 14 sec | 22 sec |
| 4K points | 3 sec | 7 sec | 11 sec |
| 2K points | 1.5 sec | 3.5 sec | 5.5 sec |

### A.2.10 Trigger Characteristics

- Trigger input
  Input level: TTL compatible
  Slope: rising or falling, selectable

  Pulse width: >100 ns
  Input impedance: 10 k, DC coupled
  Latency:
    Burst: <100 ns (typical)
    Sweep: <10 μs (typical)
  Jitter (rms)
    Burst: 1 ns; except pulse, 300 ps
    Sweep: 2.5 μs

- Trigger output
  Level: TTL compatible into 50
  Pulse width: >450 ns
  Maximum rate: 1 MHz
  Fanout: 4 HP33250As

### A.2.11 Clock Reference

- Phase Offset
  Range: –360° to +360°
  Resolution: 0.001°
- External reference input
  Lock range: 10 MHz ± 35 kHz
  Level: 100 mVpp to 5 Vpp
  Impedance: 1 knominal, AC coupled
  Lock time: <2 s
- Internal reference output
  Frequency: 10 MHz
  Level: 632 mV pp (0 dbm), nominal
  Impedance: 50 nominal, AC coupled

### A.2.12 Sync Output

Level: TTL compatible into > 1 k
Impedance: 50 nominal

### A.2.13 General Specifications

Power supply: 100–240 V, 50–60 Hz or 100–127 V, 50–400 Hz
Power consumption: 140 VA
Operating temp: 0°C to 55°C

Storage temp: –30°C to 70°C

Stored states: 4 named user configurations

Power on state: default or last

Interface: IEEE-488 and RS-232 std.

Language: SCPI-1997, IEEE-488.2

Dimensions (w × h × d)

Bench top: 254 × 104 × 374 mm

Rack mount: 213 × 89 × 348 mm

Weight: 4.6 kg

Safety designed to EN61010-1, CSA1010.1, UL-311-1

EMC tested to EN55011, IEC-1326-1

Vibration and shock: MIL-T-28800E, Type III, Class 5

Acoustic noise: 40 dBA

Warm-up time: 1 hour

Calibration interval: 1 year

Warranty: 1 year

## A.3 Operating Instructions for HP 3324A Synthesized Function/Sweep Generator

The front panel of the HP 3324A has several operating keys to control it in either the single frequency or sweep mode operation. The experiments in this book involve only the single frequency operation, and hence, only this operation will be discussed here. For further details, please refer to the Hewlett Packard 3324A Manual or the Agilent Web site.[1]

Following is the sequence of steps to be performed while operating the HP 3324A Synthesized Function/Sweep generator.

- Turn on the HP 3324A by pressing the power switch **standby**. Power is then applied to all of the HP 3324A circuits, and self tests are performed automatically by the instrument.
- Check that the **signal on/off** key is *not* lit. It is an important precaution to set the specifications of the waveform before changing the **signal on/off** key to lit position.
- Press the **function** key to select the waveform that is required. Use the arrow keys to move up and down the menu. Examples of waveforms are **sine, square, triangular**. Press **select** after highlighting the required waveform.

- Press the **frequency** key to set the frequency of the waveform. For example, use the numeric keys to select 1 MHz as the frequency of the waveform.
- Press the **amplitude** key to set the peak-to-peak amplitude of the waveform. For example, set the amplitude at 1 volt.
- Set the **phase** key at 0 degrees, unless other values are specified.
- Set the **dc offset** at 0 volt, unless specified otherwise.
- After selecting and setting the values of the waveform, press the **signal on/off** key so that it is in the lit position. The signal can be fed out the BNC connector to the circuit.

## Reference

1. Agilent Technologies, *Agilent 33250A Function/Arbitrary Waveforme Generator,* http://cp.literature.agilent.com/litweb/pdf/5968-8807EN.pdf

# APPENDIX B

# HP/Agilent 8590L RF Spectrum Analyzer

## B.1 Introduction

The HP 8590L[1] is a low-cost, but full-featured, frequency-accurate RF spectrum analyzer designed to meet general purpose measurement needs. The easy-to-use interface provides access to more than 200 built-in functions. Some of the key features of the HP8590L Spectrum Analyzer, as shown in Figure B.1, are the following:

- **Frequency counter:** Eliminates the need for a separate frequency counter with the built-in frequency counter, with ±2.1 kHz accuracy at 1 GHz (±7.6 kHz from 0 to 50°C).
- **Multiple resolution bandwidth filters:** Optimizes the tradeoffs of speed, sensitivity, and the separation of closely spaced signals with the user's choice of 10 resolution bandwidth filters, beginning at 1 kHz.
- **145 dB amplitude measurement range:** Measure signals directly with –115 dBm to +30 dBm amplitude measurement range.
- **One-button measurement routines:** Saves time, setup, and training with one button measurement routines, such as adjacent channel power, signal bandwidth, and third order intercept (TOI).
- **Phase noise of 105 dBc/Hz at 30 kHz offset:** Uncovers small signals close to carriers with an internal phase noise.
- **Dual interfaces:** Enables user to operate remotely and print directly with the optional dual interfaces that combine either an HP-IB or RS-232 port with a parallel (Centronics) port.
- **Built-in tracking generator:** Measures the scalar characteristics of your components with the optional built-in tracking generator and the HP 85714A scalar measurements personality.

---

Some material in this Appendix is reproduced with permission from Agilent Technologies Inc., Palo Alto, CA.

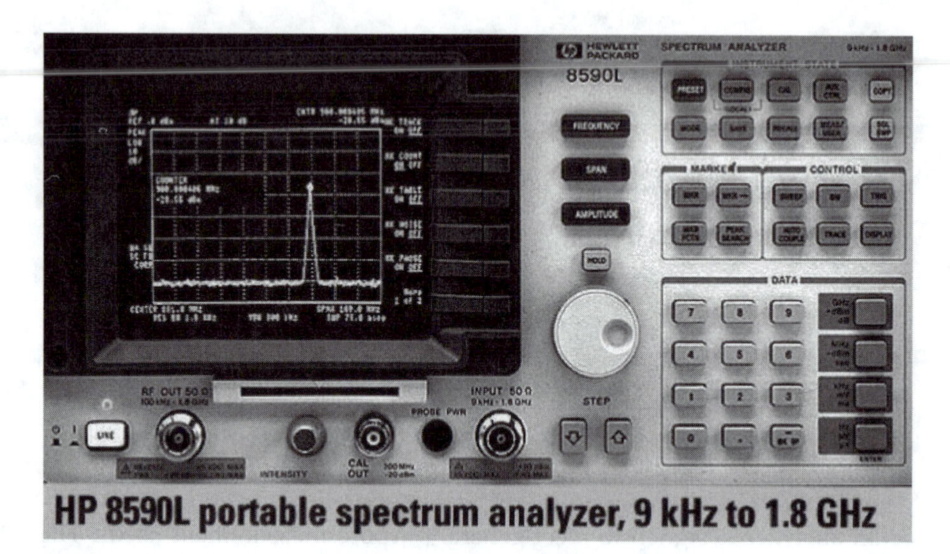


**FIGURE B.1**
HP 8590L Spectrum Analyzer. (Courtesy of Agilent Technologies Inc., Palo Alto, CA.)

## B.2 Technical Specifications

### B.2.1 Frequency Specifications

- Frequency range

  50 ohm: 9 kHz to 1.8 GHz

  75 ohm (Opt001): 1 MHz to 1.8 GHz
- Frequency readout accuracy

  (Start, stop, center, marker): ± (frequency readout × freq ref error + span accuracy + 1% of span + 20% of RBW + 100 Hz)
- Marker frequency counter accuracy

  Span 10 MHz: ± (marker freq x freq ref error + counter resolution + 1 kHz)
- Counter resolution

  Span 10 MHz: Selectable from 100 Hz to 100 kHz
- Frequency span

  Range: 0 Hz (zero span), 10 kHz to 1.8 GHz

  Resolution: Four digits or 20 Hz, whichever is greater

  Accuracy Span 10 MHz: ±3% of span

- Sweep time

  Range, span = 0 Hz or >10 kHz: 20 ms to 100 s

  Accuracy: 20 ms to 100 s: ±3%

  Sweep trigger: Free run, single, line, video, external

### B.2.2 Bandwidth Filters

- Resolution bandwidths

  1 kHz to 3 MHz (3 dB) in 1, 3, 10 sequence; 9 kHz and 120 kHz (6 dB) EMI bandwidths.

  Accuracy: ±20%

  Selectivity (characteristic) –60 dB/–3 dB:

  3 to 10 kHz: 15:1

  100 kHz to 3 MHz: 15:1

  1 kHz, 30 kHz: 16:1

- Video bandwidth range

  30 Hz to 1 MHz in 1, 3, 10 sequence

- Stability

  Noise sidebands (1 kHz RBW, 30 Hz VBW, sample detector) >10 kHz offset from CW signal: –90 dBc/Hz

  >20 kHz offset from CW signal: –100 dBc/Hz

  >30 kHz offset from CW signal: –105 dBc/Hz

### B.2.3 Amplitude Specifications

- Measurement range

  Displayed average noise level to +30 dBm

  Opt 001: Displayed average noise level to +75 dBmV

- Maximum safe input (input attenuator >=10 dB)

| | | |
|---|---|---|
| Average continuous power: | | +30 dBm (1 W) |
| | Opt 001: | +75 dBmV (0.4 W) |
| Peak pulse power: | | +30 dBm (1 W) |
| | Opt 001: | +75 dBmV (0.4 W) |
| DC: | | 25 Vdc |
| | Opt 001: | 100 Vdc |

- Gain compression

  (>10 MHz):

- Spurious responses

  Second harmonic distortion:

  5 MHz to 1.8 GHz: <–70 dBc for –45 dBm tone at input mixer

- Residual responses (input terminated, 0 dB attenuation)

  150 kHz to 1.8 GHz: <–90 dBm

- Frequency response (10 dB input attenuation)

  Absolute (referenced to 300 MHz CAL OUT): ±1.5 dB

  Relative (referenced to midpoint between highest and lowest frequency response deviations): ±1.0 dB

- Calibrator output

  Amplitude: –20 dBm ± 0.4 dB

  Opt 001: +28.75 dBmV ± 0.4 dB

- Resolution bandwidth switching uncertainty (reference to 3 kHz RBW, at reference level)

  3 kHz to 3 MHz RBW: ± 0.4 dB

  1 kHz RBW: ± 0.5 dB

- Log to linear switching

  ±0.25 dB at reference level

- Display scale fidelity

  Log incremental accuracy (0 to –60 dB from reference level): ±0.4 dB/4 dB

  Log maximum cumulative (0 to –70 dB from reference level): ±(0.4 + 0.01 × dBfrom reference level)

  Linear accuracy: ±3% of reference level

### B.2.4 General Specifications

- Environmental

  MIL-T-28800: Has been type-tested to the environmental specifications of MIL-T-28800 Class 5

  Temperature

  Operating: 0°C to +55°C

  Storage: –40°C to +75°C

  EMI compatibility:

  Conducted and radiated interference

  CISPR Pub.11/1990 Group 1 Class A

  Audible noise: Power requirements on (line 1):

  90 to 132 V rms, 47 to 440 Hz

  195 to 250 V rms, 47 to 66 Hz

- Power consumption memory
  User program memory (nominal): 238 KB nonvolatile RAM
- Data storage (nominal)
  Internal: 50 traces, 8 states
  External memory cards:
  - HP 85700A (32 KB), 24 traces or 32 states
  - HP 85702A (128 KB), 99 traces or 128 states
- Video cassette recorder
  Continuous video recording of display supported through composite video output
- Size (nominal, without handle, feet, or front cover)
  325 mm W × 163 mm H × 427mm D
- Weight
  15.2 kg

### B.2.5 System Options

Option 010 and 011 built-in tracking generators

- Frequency range
  - Opt 010: 100 kHz to 1.8 GHz
  - Opt 011: 1 MHz to 1.8 GHz
- Output level range
  - Opt 010: –15 dBm to 0 dBm
  - Opt 011: +27.8 to +42.8 dBmV
- Resolution: 0.1 dB
  - Absolute accuracy at 300 MHz, –10 dBm (+38.8 dBm V, Opt 011): ±1.5 dB

  Vernier range: 15 dB
  Accuracy: ±1 dB
  Output flatness: ±1.75 dB
  Spurious output
  - Harmonic spurs: 0 dBm (+42.8 dBmV, Opt 011) output
- Dynamic range (characteristic, max output level – TG feedthrough)
  - Opt 010: 106 dB
  - Opt 011: 100 dB
- Power sweep range
  - Opt 010: –15 dBm to 0 dBm
  - Opt 011: +28.7 to +42.8 dBmV
- Resolution: 0.1 dB

### B.2.6 General Options

Opt 003 Memory card reader
Opt 015 Protective soft tan carrying/operating case
Opt 016 Protective soft yellow carrying/operating case
Opt 040 Front panel protective cover with storage and CRT sun shield
Opt 041 HP-IB and parallel printer interfaces
Opt 042 Protective soft carrying case/backpack
Opt 043 RS-232 and parallel printer interfaces
Opt 908 Rack mount kit without handles
Opt 909 Rack mount kit with handles

- Component test
  Opt 010 Tracking generator (100 kHz to 1.8 GHz)
- Cable TV
  Opt 001 75 ohm Input
  Opt 011 Tracking generator (75 ohm, 1 MHz to 1.8 GHz)
  Opt 711 50/75 ohm matching pad/100 Vdc block
- Warranty and support
  Opt 0Q8 Factory service training
  Opt UK6 Commercial calibration certificate with test data
  Opt AB* Quick reference guide in local languages
  Opt W30 Two additional years return-to-HP service
  Opt W32 Two additional years return-to-HP calibration
  Opt 915 Component level information and service guide
- Application measurement cards/personalities
  (Requires Opt 003 memory/measurement card reader)
  HP 85700A Blank 32-KB memory card
  HP 85702A Blank 128-KB memory card
  HP 85714A Scalar measurement personality
  HP 85721A Cable TV measurement personality
  HP 85921A Cable TV PC software for HP 85721A

## B.3 Operating Principle of HP 8590L RF Spectrum Analyzer

The HP 8590L Spectrum Analyzer is a very powerful measuring tool in signal processing and communications. The instrument can very accurately measure the frequency spectrum of input signals in the frequency range of 100 KHz to 1.8 GHz. Basically, the hardware in the instrument very rapidly calculates the Fourier spectrum of the input signal using the Fast Fourier Transform (FFT).

The front panel of the HP 8590L has three primary operating keys — **frequency**, **span**, and **amplitude**. The following is the sequence of steps to be performed while operating the HP 8590L Synthesized Function/Sweep Generator. Please refer also to the HP 8590L information website [1].

- Turn on the HP 8590L by pressing the power switch **line**. Power is then applied to all of the HP 8590L circuits, and self tests are performed automatically by the instrument.
- Press the **preset** key to automatically calibrate the instrument before starting measurements.
- Press the **frequency** key. Set the desired frequency range in either of the following two ways: Set the **start** and **stop** frequencies, or set the **center** frequency and enter the **span** of frequency. For example, enter **start** as 0.5 GHz, **stop** as 1.5 GHz, or enter **center** frequency as 1 GHz, with a **span** of 1 GHz.
- The **amplitude** key is used to set the amount of attenuation required on input signal, if it is too strong. Note that signals greater than 30 dBm [Power, dBm = 10 Log (Power, mW)] should not be fed into the instrument, otherwise it will damage the analyzer.
- If there is an input signal present, the spectrum of the signal will be displayed on the screen. In order to facilitate measurements, the following controls can be used:
    - **BW (bandwidth):** Press the **bandwidth** key, and lower or increase the **resolution bandwidth** or the **video bandwidth** to improve the clarity of the displayed spectrum.
    - **MARKER:** Press the **marker** key and an option menu is displayed. Press **peak search** and the instrument will automatically track the peak value of the spectrum and position the marker cursor on the peak. Press **next peak right** or **next peak left** to move the marker to the adjoining right peak or left peak, respectively.

## Reference

1. Agilent Technologies, *Agilent 8590L-Series Portable Spectrum Analyzers, Fact Sheet,* http://cp.literature.agilent.com/litweb/pdf/5962-7275E.pdf

# Appendix C

## HP/Agilent 35665A/35670A Dynamic Signal Analyzers

### C.1 Introduction

The HP 35670A,[1] which is the successor of the HP 35665A, is a versatile Dynamic Signal Analyzer with a built-in source for general spectrum and network analysis and for octave, order, and correlation analysis. Rugged and portable, it is ideal for field work, yet it has the performance and functionality required for demanding R&D applications.

The built-in source, with optional analysis features, optimizes the instrument for analyzing and troubleshooting noise, vibration, and acoustic problems, evaluating and solving rotating machinery problems, and characterizing control systems parameters. Some of the important features of the Agilent 35670 are the following:

- Frequency range of 102.4 kHz at 1 channel, 51.2 kHz at 2 channel, 25.6 kHz at 4 channel
- 100, 200, 400, 800, and 1600 lines of resolution
- 90 dB dynamic range, 130 dB in swept-sine mode
- Source: random, burst random, periodic chirp, burst chirp, pink noise, sine, arbitrary waveform
- Measurements: linear, cross, and power spectrum, power spectral density, frequency response, coherence, THD, harmonic power, time waveform, auto-correlation, cross-correlation, histogram, PDF, CDF
- Octave analysis with triggered waterfall display
- Tachometer input and order tracking with orbit diagram
- Built-in 3.5-inch floppy disk drive

Photographs of the earlier generation model, HP 35665A, and the current model, HP 35670A, are shown in Figure C.1 and Figure C.2.

---

Some material in this Appendix is reproduced with permission from Agilent Technologies Inc., Palo Alto, CA.

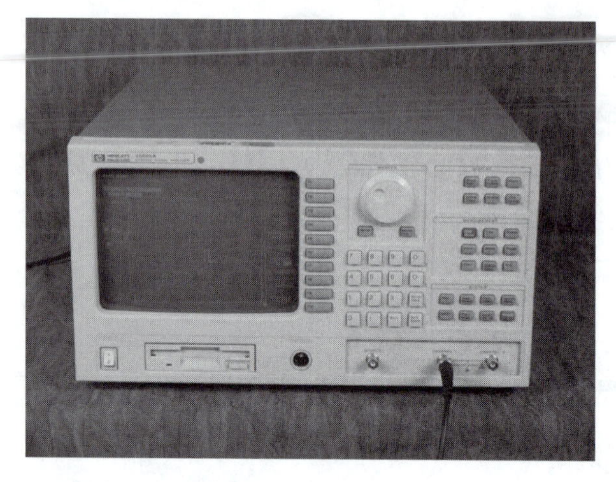

**FIGURE C.1**
Photograph of the HP 35665A Dynamic Signal Analyzer. (Courtesy of Agilent Technologies Inc., Palo Alto, CA.)

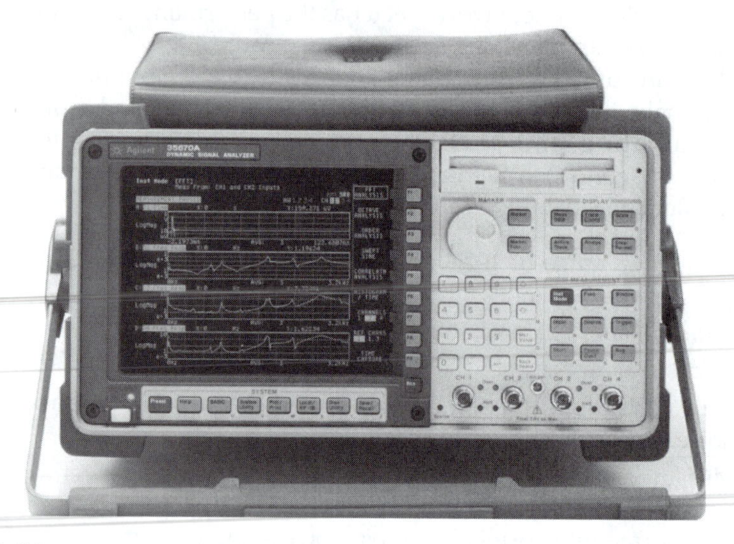


**FIGURE C.2**
Photograph of the HP 35670A Dynamic Signal Analyzer. (Courtesy of Agilent Technologies Inc., Palo Alto, CA.)

## C.2 Technical Specifications of the Agilent HP35670A

*Note:* Instrument specifications apply after 15 minutes warm-up, and within 2 hours of the last self-calibration. When the internal cooling fan has been turned off, specifications apply within 5 minutes of the last self-calibration.

All specifications are with 400-line frequency resolution and with anti-alias filters enabled unless stated otherwise.

### C.2.1 Frequency Specifications

- Maximum range
  1 channel mode: 102.4 kHz, 51.2 kHz (option AY6*)
  2 channel mode: 51.2 kHz
  4 channel mode (option AY6 only): 25.6 kHz
- Spans
  1 channel mode: 195.3 mHz to 102.4 kHz
  2 channel mode: 97.7 mHz to 51.2 kHz
  4 channel mode (option AY6 only): 97.7 mHz to 25.6 kHz
- Minimum resolution
  1 channel mode: 122 μHz (1600 line display)
  2 channel mode: 61 μHz (1600 line display)
  4 channel mode (option AY6 only): 122 μHz (800 line display)
- Maximum real-time bandwidth
  FFT span for continuous data acquisition (preset, fast averaging):
  1 channel mode: 25.6 kHz
  2 channel mode: 12.8 kHz
  4 channel mode (option AY6 only): 6.4 kHz
- Measurement rate
  Typical (preset, fast averaging):
  1 channel mode: 70 averages/second
  2 channel mode: 33 averages/second
  4 channel mode (option AY6 only): 15 Averages/Second
- Display update rate
  Typical (preset, fast average OFF): 5 updates/second
  Maximum: 9 updates/second
  (Preset, fast average off, single channel, single display, undisplayed trace displays set to data registers)
- Accuracy
  ±30 ppm (.003%)

### C.2.2 Single Channel Amplitude

Absolute amplitude accuracy (FFT)

(A combination of full scale accuracy, full scale flatness, and amplitude linearity)

±2.92% (0.25 dB) of reading
±0.025% of full scale

- FFT full scale accuracy at 1 kHz (0 dBfs)
  ±0.15 dB (1.74%)
- FFT full scale flatness (0 dBfs), relative to 1 kHz
  ±0.2 dB (2.33%)
- FFT amplitude linearity at 1 kHz
  Measured on +27 dBVrms range with time averaging
  0 to –80 dBfs ±0.58% (0.05 dB) of reading ±0.025% of full scale
- Amplitude resolution
  16 bits less 2 dB over-range with averaging 0.0019% of full scale (typical)
- Residual DC response (FFT mode)
  Frequency display (excludes A-weight filter)
  <–30 dBfs or 0.5 mVdc

### C.2.3 FFT Dynamic Range

- Spurious free dynamic range
  90 dB typical (<–80 dBfs)
  (Includes spurs, harmonic distortion, intermodulation distortion, alias products)
  Excludes alias responses at extremes of span
  Source impedance = 50
  800 line display
- Full span FFT noise floor (typical)
  Flat top window, 64 RMS averages, 800 line display
- Harmonic distortion
  <–80 dBfs
  Single tone (in band), 0 dBfs
- Intermodulation distortion
  <–80 dBfs
  Two tones (in-band), each –6.02 dBfs
- Spurious and residual responses
  <–80 dBfs
  Source impedance = 50
- Frequency alias responses
  Single tone (out of displayed range), 0 dBfs, 1 MHz
  (200 kHz with IEPE transducer power supply on)

2.5% to 97.5% of the frequency span: <–80 dBfs

Lower and upper 2.5% of frequency span: <–65 dBfs

### C.2.4 Input Noise

- Input noise level

  Flat top window, –51 dBVrms range, source impedance = 50 Ω

  Above 1280 Hz: <–140 dBVrms/vHz

  160 Hz to 1280 Hz: <–130 dBVrms/vHz

  *Note:* To calculate noise as dB below full scale:

  Noise [dBfs] = noise [dB/vHz] + 10 LOG (NBW) – range [dBVrms]; where NBW is the noise equivalent BW of the window (see below)

### C.2.5 Window Parameters

| | Uniform | Hann | Flat Top |
|---|---|---|---|
| –3 dB Bandwidth | 0.125% of span | 0.185% of span | 0.450% of span |
| Noise Equivalent: Bandwidth | 0.125% of span | 0.1875% of span | 0.4775% of span |
| Attenuation at ±1/2 Bin: | 4.0 dB | 1.5 dB | 0.01 dB |
| Shape Factor (–60 dB BW/–3 dB BW) 800 Hz Span | 716 | 9.1 | 2.6 |

### C.2.6 Single Channel Phase

- Phase accuracy relative to external trigger: ±4.0 deg

  16 time averages center of bin, DC coupled

  0 dBfs to –50 dBfs only

  0 Hz < freq = 10.24 kHz only

  For Hann and flat top windows, phase is relative to a cosine wave at the center of the time record. For the uniform, force, and exponential windows, phase is relative to a cosine wave at the beginning of the time record.

### C.2.7 Cross-Channel Amplitude

- FFT cross-channel gain accuracy : ± 0.04 dB (0.46%)

  Frequency response mode, same amplitude range

  At full scale: tested with 10 RMS averages on the –11 to +27 dBVrms ranges, and 100 RMS averages on the –51 dBVrms range

### C.2.8 Cross-Channel Phase

- Cross-channel phase accuracy: ±0.5 deg
  (Same conditions as cross-channel amplitude)

### C.2.9 Input

- Input ranges (full scale)
  Auto-range capability: +27 dBVrms (31.7 Vpk) to –51 dBVrms (3.99 mVpk) in 2 dB steps
- Maximum input levels
  42 Vpk
- Input impedance
  1 MΩ ±10%, 90 μF nominal
- Low side to chassis impedance
  1 MΩ ±30% (typical)
  Floating mode: <0.010 μF
  Grounded mode: 100. Ω
- AC coupling rolloff
  <3 dB rolloff at 1Hz
  Source Impedance = 50
- Common mode rejection ratio
  Single tone at or below 1 kHz
  –51 dBVrms to –11 dBVrms ranges: >75 dB typical
  –9 dBVrms to +9 dBVrms ranges: >60 dB typical
  +11 dBVrms to +27 dBVrms ranges: >50 dB typical
- Common mode range (floating mode)
  ± 4V pk
- IEPE transducer power supply
  Current source: 4.25 ± 1.5 mA
  Open circuit voltage: +26 to +32 Vdc
- A-weight filter, type 0 tolerance
  Conforms to ANSI standard S1.4-1983; and to IEC 651-1979; 10 Hz to 25.6 kHz
- Crosstalk
  Between input channels and source-to-input (receiving channel source impedance = 50 Ω̃): <–135 dB below signal or <–80 dBfs of receiving channel, whichever response is greater in amplitude

- Time domain
  Specifications apply in histogram/time mode and unfiltered time display
- DC amplitude accuracy
  ±5.0% fs
- Rise time of –1V to 0V test pulse
  <11.4 μSec
- Settling time of –1V to 0V test pulse
  <16 μSec to 1%
- Peak overshoot of –1V to 0V test pulse
  <3%
- Sampling period
  1 channel mode: 3.815 μsec to 2 sec. in 2x steps
  2 channel mode: 7.629 μsec to 4 sec. in 2x steps
  4 channel mode (option AY6 only): 15.26 μsec. to 8 sec in 2x steps

### C.2.10 Trigger

- Trigger modes
  Internal, source, external (analog setting) GPIB
- Maximum trigger delay
  Post trigger: 8191 seconds
  Pre trigger: 8191 sample periods
  No two channels can be farther than ±7168 samples from each other.
- External trigger max input
  ±42 Vpk
- External trigger range
  Low range: –2V to +2V
  High range: –10V to +10V
- External trigger resolution
  Low range: 15.7 mV
  High range: 78 mV

### C.2.11 Tachometer

- Pulses per revolution
  0.5 to 2048

- RPM
  5 RPM 491,519
- RPM accuracy
  ±100 ppm (0.01%) (typical)
- Tach level range
  Low range: –4V to +4V
  High range: –20V to +20V
- Tach level resolution
  Low range: 39 mV
  High range: 197 mV
- Maximum tach input level
  ±42 Vpk
- Minimum tach pulse width
  600 nSec
- Maximum tach pulse rate
  400 kHz (typical)

### C.2.12 Source Output

- Source types
  Sine, random noise, chirp, pink noise, burst, random, burst chirp
  Amplitude range (Vacpk + |Vdc| = 10V)
  AC: ±5V peak
  DC: ±10V
- AC amplitude resolution
  Voltage > 0.2 Vrms: 2.5 mVpeak
  Voltage < 0.2 Vrms: 0.25 mVpeak
- DC offset accuracy
  ±15 mV ± 3% of (|DC| + Vacpk) settings
- Pink noise adder
  Add 600 mV typical when using pink noise
- Output impedance
  <5̃ Ω
- Maximum loading
  Current: ±20 mA peak
  Capacitance: 0.01 μF
- Sine amplitude accuracy at 1 kHz: ±4% (0.34 dB) of setting
  Rload > 250̃ Ω, 0.1 Vpk to 5 Vpk

- Sine Flatness
  Relative to 1 kHz: ±1 dB
  0.1V to 5V peak
- Harmonic and subharmonic distortion and spurious signals (in band)
  0.1 Vpk to 5 Vpk Sine Wave
  Fundamental < 30 kHz: <–60 dBc
  Fundamental > 30 kHz: <–40 dBc

### C.2.13 Digital Interfaces

- External keyboard
  Compatible with PC-style 101-key keyboard
- GPIB conforms to the following standards:
  IEEE 488.1 (SH1, AH1, T6, TE0, L4, LE0, SR1, RL1, PP0, DC1, DT1, C1, C2, C3, C12, E2), IEEE 488.2-1987; complies with SCPI 1992
- Data transfer rate
  <45 mSec for a 401 point trace
  (REAL 64 Format)
- Serial port
- Parallel port
- External VGA port

### C.2.14 Computed Order Tracking — Option 1D0

$$\frac{\text{Maximum Order} \times \text{Maximum RPM}}{60} =$$

- Online (real time)
  1 channel mode: 25,600 Hz
  2 channel mode: 12,800 Hz
  4 channel mode: 6,400 Hz
- Capture playback
  1 channel mode: 102,400 Hz
  2 channel mode: 51,200 Hz
  4 channel mode: 25,600 Hz
- Number of orders 200
  $5 \leq \text{RPM} \leq 491{,}519$
  (Maximum useable RPM is limited by resolution, tach pulse rate, pulses/revolution, and average mode settings.)

- Delta order
  1/128 to 1/1
- Resolution
  ≤400
  (Maximum order)/(Delta order)
- Maximum RPM ramp rate: 1000 RPM/second real-time (typical)
  1000 – 10,000 RPM run up
  Maximum order: 10
  Delta order: 0.1
  RPM step: 30 (1 channel)
  60 (2 channel)
  120 (4 channel)
- Order track amplitude accuracy
  ±1 dB (typical)

### C.2.15 Real Time Octave Analysis — Option 1D1

- Standards
  Conforms to ANSI standard S1.11 – 1986, order 3, type 1-D, Extended and Optional Frequency Ranges
  Conforms to IEC 651-1979 type 0 Impulse, and ANSI S1.4
- Frequency ranges (at centers)

| | 1 Channel | 2 Channel | 4 Channel |
|---|---|---|---|
| Online (real time): | | | |
| 1/1 Octave | 0.063–16 kHz | 0.063–8 kHz | 0.063–4 kHz |
| 1/3 Octave | 0.08–40 kHz | 0.08–20 kHz | 0.08–10 kHz |
| 1/12 Octave | 0.0997–12.338 kHz | 0.0997–6.169 kHz | 0.0997–3.084 kHz |
| Capture playback: | | | |
| 1/1 Octave | 0.063–16 kHz | 0.063–16 kHz | 0.063–16 kHz |
| 1/3 Octave | 0.08–31.5 kHz | 0.08–31.5 kHz | 0.08–31.5 kHz |
| 1/12 Octave | 0.0997–49.35 kHz | 0.0997–49.35 kHz | 0.0997–49.35 kHz |

One to 12 octaves can be measured and displayed.

1/1-, 1/3-, and 1/12-octave true center frequencies related by the formula: f(i+1)/f(i) = 2^(1/n); n=1, 3, or 12; where 1000 Hz is the reference for 1/1, 1/3 octave, and 1000*2^(1/24) Hz is the reference for 1/12 octave. The marker returns the ANSI standard preferred frequencies.

- Accuracy
  1 second stable average
  Single tone at band center: ±0.20 dB
  Readings are taken from the linear total power spectrum bin.
- 1/3-octave dynamic range*:* >80 dB (typical) per ANSI S1.11-1986
  2 sec. stable average
  Total power limited by input noise level

### C.2.16 Swept Sine Measurements — Option 1D2

- Dynamic range: 130 dB
  Tested with 11 dBVrms
  Source level at: 100 msec integration

### C.2.17 Arbitrary Waveform Source — Option 1D4

- Amplitude range (Vacpk + $|\text{Vdc}|$ = 10V)
  AC: ±5V peak
  DC: ±10V
- Record length
  # of points = 2.56 × lines of resolution,
  # of complex points = 1.28 × lines of resolution
- DAC resolution
  0.2828 Vpk to 5 Vpk: 2.5 mV
  0 Vpk to 0.2828 Vpk: 0.25 mV

## C.3 General Specifications

- Safety standards CSA certified for electronic test and measurement equipment per CSA C22.2, NO. 231
  This product is designed for compliance to: UL1244, Fourth Edition
  IEC 348, Second Edition, 1978
- EMI/RFI standards
  CISPR 11

- Acoustic power

  LpA < 55 dB (cooling fan at high speed setting)

  < 45 dB (auto speed setting at 25°C)

  Fan speed settings of high, automatic, and off are available. The fan off setting can be enabled for a short period of time, except at higher ambient temperatures where the fan will stay on.

- Environmental operating restrictions

| | **Operating (Disk in drive)** | **Operating (No Disk in drive)** | **Storage and Transport** |
|---|---|---|---|
| Ambient temp. | 4°C–45°C | 0°C–55°C | –40°C–70°C |
| Relative humidity (noncondensing) | | | |
| Minimum | 20% | 15% | 5% |
| Maximum | 80% at 32°C | 95% at 40°C | 95% at 50°C |
| Vibrations (5–500 Hz) | 0.6 Grms | 1.5 Grms | 3.41 Grms |
| Shock | 5G (10 mSec 1/2 sine) | 5G (10 mSec 1/2 sine) | 40G (3 mSec 1/2 sine) |
| Max. altitude | 4600 meters (15,000 ft) | 4600 meters (15,000 ft) | 4600 meters (15,000 ft) |

- AC power

  90 Vrms – 264 Vrms, (47–440 Hz)

  350 VA maximum

- DC power

  12 VDC to 28 VDC nominal

  200 VA maximum

- DC current at 12V

  Standard: <10A (typical)

  4 channel: <12A (typical)

- Warm-up time

  15 minutes

- Weight

  15 kg (33 lb) net

  29 kg (64 lb) shipping

- Dimensions (excluding bail handle and impact cover)

  Height: 190 mm (7.5″)

  Width: 340 mm (13.4″)

  Depth: 465 mm (18.3″)

## C.4 Operating Principle of HP 35665A Dynamic Signal Analyzer

The HP 35665A Dynamic Signal Analyzer is a very advanced and versatile signal measuring equipment. The instrument can work in two modes and perform a variety of functions. The HP 35665A can handle baseband signals from 0 to 100 KHz. and has a very high resolution in that range. The two operating modes are **single channel** and **dual channel** modes. In the single channel mode, the instrument acts as a time/frequency measuring tool, whereas in the dual channel mode, the instrument can measure the frequency response of a circuit device, such as a filter.

### C.4.1 Single Channel Mode Operation

- Turn on the HP 3324A by pressing the power key. Press the **preset** key and press **do preset**, which calibrates the instrument back to its default values.
- Press **inst mode** and press **1 channel**. Now the instrument is set to perform time/frequency measurements on input signals.
- Connect the input signal to either **channel 1** or **channel 2**.
- Press **meas data**, and select either **time (channel 1 or 2)** or **spectrum (channel 1 or 2)**.
- Press **autoscale** and the time or frequency graph will be displayed. Use the **marker** and **marker to peak** keys to determine the amplitude levels of the signal.

### C.4.2 Dual Channel Mode Operation

- Turn on the HP 35665A by pressing the power key. Press the **preset** key and press **do preset**, which calibrates the instrument back to its default values.
- Press **inst mode** and press **2 channel**. Now the instrument is set to perform device frequency response measurements.
- Connect the **source** key of the HP 35665A to the input port of the Device Under Test (DUT) using BNC cable. Using a BNC Tee, simultaneously connect the **source** key of the HP 35665A to **channel 1**.
- Connect the output port of the DUT to **channel 2** of the HP 35665A. Now all connections are complete for making frequency response measurements.

- Press **meas data** and press **frequency response**.
- Press **source** and toggle to **source on/off**. Select any one of the different sources listed, for example, **random noise**.
- Press **level** and set the voltage (or power) level of the source.
- Press **autoscale** and the frequency response of the DUT will now be displayed.
- Use the **marker** and **marker to peak** keys to determine the amplitude levels of the signal.

## Reference

1. http://cp.literature.agilent.com/litweb/pdf/5966-3064E.pdf

# Appendix D

## *HP/Agilent 54500/54600 Series Digitizing Oscilloscopes*

### D.1 Introduction

The Agilent 54624A 100 MHz Oscilloscope (one in the Agilent 54600 series of oscilloscopes) provides the channel count and measurement power that the user needs, including MegaZoom deep memory, high definition display, and flexible triggering, especially if designs include heavy analog content. Whether testing is for designs with four inputs, such as antilock brakes, or monitoring multiple outputs of a power supply, the four-channel model helps you get your debug and verification done with ease. Some of the important features of this equipment are as follows:

- Enhanced serial triggering capabilities and integrated 5-digit frequency counter measurement
- Lower cost deep memory 4-channel scope on the market
- Unique 4-channel model
- 100 MHz, 200 MSa/sec.
- 2 MB of MegaZoom deep memory per channel
- Patented high-definition display
- Flexible triggering including I²C, SPI, CAN, and USB

Figure D.1 shows the photograph of the HP 54501A Digitizing Oscilloscope, and Figure D.2 shows the photograph of the newer model, HP 54624A.

Some material in this Appendix is reproduced with permission from Agilent Technologies Inc., Palo Alto, CA.

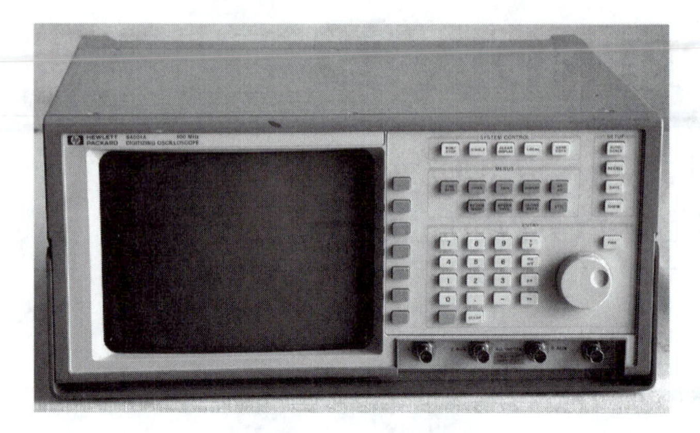

**FIGURE D.1**
HP 54501A 100 MHz Digitizing Oscilloscope. (Courtesy of Agilent Technologies Inc., Palo Alto, CA.)

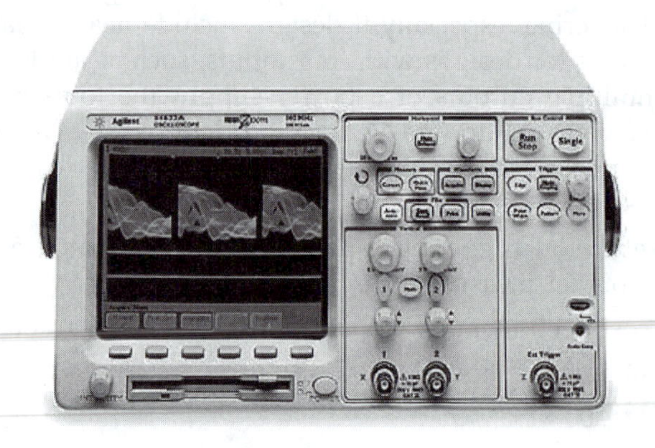

**FIGURE D.2**
HP 54624A 100 MHz Digitizing Oscilloscope. (Courtesy of Agilent Technologies Inc., Palo Alto, CA.)

## D.2 Performance Characteristics of the Agilent 54600 Series Digitizing Oscilloscopes

### D.2.1 Acquisition: Analog Channels

- Max sample rate
  54621A/D, 54622A/D, 54624A: 200 MSa/s
  54641A/D, 54642A/D: 2 GSa/sec interleaved, 1 GSa/sec. each channel

- Max memory depth
  54621A/D, 54622A/D, 54624A: 4 MB interleaved, 2 MB each channel
  54641A/D, 54642A/D: 8 MB interleaved, 4 MB each channel
- Vertical resolution
  8 bits
- Peak detection
  54621A/D, 54622A/D, 54624A: 5 ns
  54641A/D, 54642A/D: 1 ns @ max sample rate
- Averaging
  Selectable from 2, 4, 8, 16, 32, 64 … to 16384
- High resolution mode
  54621A/D, 54622A/D, 54624A: 12 bits of resolution when = 500 μs/div (average mode with ave = 1)
  54641A/D, 54642A/D: 12 bits of resolution when =100 μs/div (average mode with ave = 1)
- Filter
  Sinx/x interpolation (single shot BW = sample rate/4 or bandwidth of scope, whichever is less) with vectors on.

### D.2.2 Acquisition: Digital Channels (54621D, 54622D, 54641D, and 54642D only)

- Max sample rate
  54621D, 54622D: 400 MSa/s interleaved, 200 MSa/s each channel
  54641D, 54642D: 1 GSa/s
- Max memory depth
  54621D, 54622D: 8 MB interleaved, 4 MB ea. channel
  54641D, 54642D: 4 MB
- Vertical resolution
  1 bit
- Glitch detection (min pulse width)
  5 ns

### D.2.3 Vertical System: Analog Channels

- Analog channels
  54621A/D, 54622A/D, 54641A/D, 54642A/D: Ch 1 and 2 simultaneous acquisition
  54624A: Ch 1, 2, 3, and 4 simultaneous acquisition

- Bandwidth (–3dB)*

  54621A/D: DC to 60 MHz

  54622A/D, 54624A: DC to 100 MHz

  54641A/D: DC to 350 MHz

  54642A/D: DC to 500 MHz

  * Denotes warranted specifications, all others are typical. Specifications are valid after a 30-minute warm-up period and ±10°C from firmware calibration temperature.

- AC coupled

  54621A/D: 3.5 Hz to 60 MHz

  54622A/D: 54624A: 3.5 Hz to 100 MHz

  54641A/D: 3.5 Hz to 350 MHz

  54642A/D: 3.5 Hz to 500 MHz

- Calculated rise time (= 0.35/bandwidth)

  54621A/D: ~5.8 ns

  54622A/D, 54624A: ~3.5 ns

  54641A/D: ~1.0 ns

  54642A/D: ~700 ps

- Single shot bandwidth

  54621A/D, 54622A/D, 54624A: 50 MHz

  54641A/D: 350 MHz maximum

  54642A/D: 500 MHz maximum

- Range1,2

  54621A/D, 54622A/D, 54624A: 1 mV/div to 5 V/div

  54641A/D, 54642A/D: 2 mV/div to 5 V/div

- Maximum Input

  CAT I 300 Vrms, 400 Vpk, CAT II 100 Vrms, 400 Vpk

  With 10073C/10074C 10:1 probe: CAT I 500 Vpk, CAT II 400 Vpk

  5 Vrms with 50 input

- Offset range

  54621A/D, 54622A/D, 54624A: ±5 V on ranges <10 mV/div; ±25 V on ranges 10 mV/div to 199 mV/div; ±100 V on ranges = 200 mV/div

  54641A/D, 54642A/D: ±5 V on ranges <10 mV/div; ±20 V on ranges 10 mV/div to 200 mV/div; ±75 V on ranges >200 mV/div

- Dynamic range

  Lesser of ±8 div or ±32 V from center screen

- Input resistance

  54621A/D, 54622A/D, 54624A: 1 M. ±1%

  54641A/D, 54642A/D: 1 M. ±1%, 50 selectable
- Input capacitance

  54621A/D, 54622A/D, 54624A: ~14 pF

  54641A/D, 54642A/D: ~13 pF
- Coupling

  54621A/D, 54622A/D, 54624A: AC, DC, ground

  54641A/D, 54642A/D: AC, DC
- BW limit

  54621A/D, 54622A/D, 54624A: ~20 MHz selectable

  54641A/D, 54642A/D: ~25 MHz selectable
- Channel-to-channel isolation (with channels at same V/div)

  54621A/D, 54622A/D, 54624A: DC to 20 MHz > 40 dB; 20 MHz to max bandwidth >30 dB

  54641A/D, 54642A/D: DC to max bandwidth >40 dB
- Probes

  54621A/D, 54622A/D, 54624A: 10:1 10074C shipped standard for each analog channel

  54641A/D, 54642A/D: 10:110073C shipped standard for each analog channel
- Probe ID (Agilent/HP and Tek compatible)

  Auto probe sense
- ESD tolerance

  ±2 kV
- Noise peak-to-peak

  54621A/D, 54622A/D, 54624A: 2% full scale or 1 mV, whichever is greater

  54641A/D, 54642A/D: 3% full scale or 3 mV, whichever is greater
- Common mode rejection ratio

  20 dB @ 50 MHz
- DC vertical gain accuracy

  ±2.0% full scale
- DC vertical offset accuracy

  54621A/D, 54622A/D, 54624A: <200 mV/div: ±0.1 div ±1.0 mV ±0.5% offset value; =200 mV/div: ±0.1 div ±1.0 mV ±1.5% offset value

  54641A/D, 54642A/D: 200 mV/div: ±0.1 div ±2.0 mV ±0.5% offset value; >200 mV/div: ±0.1 div ±2.0 mV ±1.5% offset value

- Single cursor accuracy

  ± {DC vertical gain accuracy + DC vertical offset accuracy + 0.2% full scale (~1/2 LSB)}

  54621A/D, 54622A/D, 54624A example: for 50 mV signal, scope set to 10 mV/div (80 mV full scale), 5 mV offset, accuracy = ±{2.0% (80mV) + 0.1 (10 mV) + 1.0 mV + 0.5% (5 mV) + 0.2% (80 mV)} = ± 3.78 mV

- Dual cursor accuracy

  ± {DC vertical gain accuracy + 0.4% full scale (~1 LSB)}

  Example: for 50 mV signal, scope set to 10 mV/div (80 mV full scale), 5 mV offset, accuracy = ± {2.0%(80 mV) + 0.4%(80 mV)} = ±1.92 mV

### D.2.4 Vertical System: Digital Channels (54621D, 54622D, 54641D, and 54642D only)

- Number of channels

  16 Digital — labeled D15 – D0

- Threshold groupings

  Pod 1: D7 – D0

  Pod 2: D15 – D8

- Threshold selections

  TTL, CMOS, ECL, user-definable (selectable by pod)

- User-defined threshold range

  ±8.0 V in 10 mV increments

- Maximum input voltage

  ±40 V peak CAT I

- Threshold accuracy

  ± (100 mV + 3% of threshold setting)

- Input dynamic range

  ±10 V about threshold

- Minimum input voltage swing

  500 mV peak-to-peak

- Input capacitance

  ~8 pF

- Input resistance

  100 k ±2% at probe tip

- Channel-to-channel skew

  2 ns typical, 3 ns maximum

### D.2.5 Horizontal

- Range

  54621A/D, 54622A/D, 54624A: 5 ns/div to 50 s/div

  54641A/D, 54642A/D: 1 ns/div to 50 s/div

- Resolution

  54621A/D, 54622A/D, 54624A: 25 ps

  54641A/D, 54642A/D: 2.5 ps

- Vernier

  1-2-5 increments when off, ~25 minor increments between major settings when on

- Reference positions

  Left, center, right

- Delay range

| | |
|---|---|
| 54621A/D, 54622A/D, 54624A: | Pre-trigger (negative delay): Greater of 1 screen width or 10 ms<br>Post-trigger (positive delay): 500 seconds |
| 54641A/D, 54642A/D: | Pre-trigger (negative delay): Greater of 1 screen width or 1 ms<br>Post-trigger (positive delay): 500 seconds |

- Analog delta-t accuracy

| | |
|---|---|
| 54621A/D, 54622A/D, 54624A: | Same channel: ±0.01% reading ±0.1% screen width ±40 ps<br>Channel-to-channel: ±0.01% reading ±0.1% screen width ±80 ps |
| 54641A/D, 54642A/D: | Same channel: ±0.005% reading ±0.1% screen width ±20 ps<br>Channel-to-channel: ±0.005% reading ±0.1% screen width ±40 ps |

  Same Channel Example (54641A/D, 54642A/D): for signal with pulse width of 10 μs, scope set to 5 μs/div (50 μs screen width), delta-t accuracy = ±{.005%(10 μs) + 0.1%(50 μs) + 20 ps} = 50.52 ns

- Digital delta-t accuracy

  54621A/D, 54622A/D, 54624A: (non-Vernier settings)

  Same channel: ±0.01% reading ±0.1% screen width ±(1 digital sample period, 2.5 or 5 ns based on sample rate of 200/400 MSa/s)

Channel-to-channel: ±0.01% reading ±0.1% screen width ±(1 digital sample period, 2.5 or 5 ns) ±chan-to-chan skew (2 ns typical, 3 ns maximum)

54641A/D, 54642A/D:

Same channel: ±0.005% reading ±0.1% screen width ±(1 digital sample period, 1ns)

Channel-to-channel: ±0.005% reading ±0.1% screen width ±(1 digital sample period) ±chan-to-chan skew

Same Channel Example (54641A/D, 54642A/D): for signal with pulse width of 10 μs, scope set to 5 μs/div (50 μs screen width), delta-t accuracy = ±{.005%(10 μs) + 0.1%(50 μs) + 1 ns} = 51.5 ns

- Delay jitter

  <1 ppm

- RMS jitter

  0.025% screen width + 30 ps

- Modes

  Main, delayed, roll, XY

- XY

  Bandwidth: max bandwidth

  Phase error @ 1 MHz: 1.8 degrees

  Z blanking: 1.4 V blanks trace (use external trigger) — 54621A/D, 54622A/D, 54624A only

### D.2.6 Trigger System

- Sources

  54621A/622A, 54641A/642A: Ch 1, 2, line, ext

  54621D/622D, 54641D/642D: Ch 1, 2, line, ext, D15 – D0

  54624A: Ch 1, 2, 3, 4, line, ext

- Modes

  Auto, triggered (normal), single auto level (54621A/D, 54622A/D, 54624A only)

- Hold off time

  ~60 ns to 10 seconds

- Selections

  Edge, pulse width, pattern, TV, duration, sequence, CAN, LIN, USB, I²C, SPI

- Edge

  Trigger on a rising or falling edge of any source

- Pattern

  Trigger on a pattern of high, low, and don't care levels and/or a rising or falling edge established across any of the sources (The analog channel's high or low level is defined by that channel's trigger level.)

- Pulse width

  Trigger when a positive- or negative-going pulse is less than, greater than, or within a specified range on any of the source channels

  Minimum pulse width setting: 5ns (2 ns on 54641A/D, 54642A/D analog channels)

  Maximum pulse width setting: 10 sec.

- TV

  Trigger on any analog channel for NTSC, PAL, PAL-M, or SECAM broadcast standards on either positive or negative composite video signals (Modes supported include Field 1, Field 2, or both, all lines, or any line within a field. Also supports triggering on noninterlaced fields. TV trigger sensitivity: 0.5 division of synch signal.)

- Sequence

  Arm on event A, trigger on event B, with option to reset on event C or time delay.

- CAN

  Trigger on CAN (Controller Area Network) version 2.0A and 2.0B signals; can trigger on the start of frame bit of a data frame, a remote transfer request frame, or an overload frame.

- LIN

  Trigger on LIN (Local Interconnect Networking) sync break at beginning of message frame.

- USB

  Trigger on USB (Universal Serial Bus) start of packet, end of packet, reset complete, enter suspend, or exit uspend on the differential USB data lines. USB low speed and full speed are supported.

- $I^2C$

  Trigger on $I^2C$ (Inter-IC bus) serial protocol at a start/stop condition or user-defined frame with address and/or data values. Also trigger on missing acknowledge, restart, EEPROM read, and 10-bit write.

- SPI

  Trigger on SPI (Serial Protocol Interface) data pattern during a specific framing period. Support positive and negative chip select framing as well as clock idle framing and user-specified number of bits per frame.

- Duration

  Trigger on a multichannel pattern whose time duration is less than a value, greater than a value, greater than a time value with a timeout value, or inside or outside of a set of time values

  Minimum duration setting: 5 ns

  Maximum duration setting: 10 s

- Autoscale

  Finds and displays all active analog and digital (for 54621D/54622D/54641D/54642D) channels, sets edge trigger mode on highest numbered channel, sets vertical sensitivity on analog channels and thresholds on digital channels, time base to display ~1.8 periods. Requires minimum voltage >10 mVpp, 0.5% duty cycle and minimum frequency >50Hz.

### D.2.7 Analog Channel Triggering

- Range (internal)

  ±6 div from center screen

- Sensitivity*

  54621A/D, 54622A/D, 54624A: greater of 0.35 div or 2.5 mV

  54641A/D, 54642A/D:

  <10mV/div: greater of 1 div or 5mV

  =10mV/div: 0.6 div

- Coupling

  AC (~3.5 Hz on 54621A/D, 54622A/D, 54624A. ~10 Hz on 54641A/D, 54642A/D), DC, noise reject, HF reject and LF reject (~50 kHz)

### D.2.8 Digital (D15 – D0) Channel Triggering (54621D, 54622D, 54641D, and 54642D)

- Threshold range (used defined)

  ±8.0 V in 10 mV increments

- Threshold accuracy

  ±(100 mV + 3% of threshold setting)

- Predefined thresholds

  TTL = 1.4 V, CMOS = 2.5 V, ECL = –1.3 V

### D.2.9 External (EXT) Triggering

- Input resistance

  54621A/D, 54622A/D, 54624A: 1 M, ±3%

  54641A/D, 54642A/D: 1 M ±3% or 50.

- Input capacitance

  54621A/D, 54622A/D, 54624A: ~14 pF

  54641A/D, 54642A/D: ~13pF
- Maximum input

  CAT I 300 Vrms, 400 Vpk; CAT II 100 Vrms, 400 Vpk

  With 10073C/10074C 10:1 probe: CAT I 500 Vpk, CAT II 400 Vpk

  5 Vrms with 50-ohm input
- Range

  54621A/D, 54622A/D, 54624A: ±10 V

  54641A/D, 54642A/D: DC coupling: trigger level ± 8V; AC coup./LFR: AC input minus trig level not to exceed ±8V
- Sensitivity

  54621A/D, 54622A/D, 54624A: DC to 25 MHz, < 75 mV; 25 MHz to max bandwidth, <150 mV

  54641A/D, 54642A/D: DC to 100 MHz, < 100 mV; 100 MHz to max bandwidth, <200 mV
- Coupling

  AC (~3.5 Hz), DC, noise reject, HF reject and LF reject (~50 kHz)
- Probe ID (Agilent/HP and Tek compatible)

  Auto probe sense for 54621A/622A/641A/642A

### D.2.10 Display System

- Display

  7-inch raster monochrome CRT
- Throughput of analog channels

  25 million vectors/sec. per channel with 32 levels of intensity
- Resolution

  255 vertical by 1000 horizontal points (waveform area) 32 levels of gray scale
- Controls

  Waveform intensity on front panel. Vectors on/off; infinite persistence on/off 8 × 10 grid with continuous intensity control
- Built-in help system

  Key-specific help in 11 languages displayed by pressing and holding key or soft key of interest
- Real-time clock

  Time and date (user settable)

### D.2.11 Measurement Features

- Automatic measurements

  Measurements are continuously updated. Cursors track current measurement.

- Voltage (analog channels only)

  Peak-to-peak, maximum, minimum, average, amplitude, top, base, overshoot, preshoot, RMS (DC)

- Time

  Frequency, period, + width, – width, and duty cycle on any channels

  Rise time, fall time, X at max (time at max volts), X at min (time at min volts), delay, and phase on analog channels only

- Counter

  Built-in 5 digit frequency counter on any channel. Counts up to 125 MHz

  Threshold definition

  Variable by percent and absolute value; 10%, 50%, 90% default for time measurements

- Cursors

  Manually or automatically placed readout of Horizontal (X,.X, 1/.X) and Vertical (Y,.Y)

  Additionally digital or analog channels can be displayed as binary or hex values

- Waveform math

  One function of 1–2, 1*2, FFT, differentiate, integrate

  Source of FFT, differentiate, integrate: analog channels 1 or 2 (or 3 or 4 for 54624A), 1–2, 1+2, 1*2

### D.2.12 FFT

- Points

  Fixed at 2048 points

- Source of FFT

  Analog channels 1 or 2 (or 3 or 4 on 54624A only), 1+2, 1–2, 1*2

- Window

  Rectangular, flattop, Hanning

- Noise floor

  –70 to –100 dB depending on averaging

- Amplitude display

  In dBV, dBm.

- Frequency resolution
  0.097656/(time per div.)
- Maximum frequency
  102.4/(time per div.)

### D.2.13 Storage

- Save/recall (nonvolatile)
  54621A/D, 54622A/D, 54624A: 3 setups and traces can be saved and recalled internally
  54641A/D, 54642A/D: 4 setups and traces can be saved and recalled internally
- Floppy disk drive
  3.5″ 1.44 MB double density
  Image formats: TIF, BMP
  Data formats: X and Y (time/voltage) values in CSV format
  Trace/setup formats: recalled

### D.2.14 I/O

- RS-232 (serial) standard port
  1 port: XON or DTR; 8 data bits; 1 stop bit; parity = none; 9600, 19200, 38400, 57600 baud rates (use Agilent 34398A cable)
- Parallel standard port
  Printer support
- Printer compatibility
  HP DeskJet, LaserJet with HP PCL 3 or greater compatibility
  Black and white @150 × 150 dpi; Gray scale @ 600 × 600 dpi
  Epson: black and white @180x180 dpi
  Seiko thermal DPU-414: black and white
- Optional GPIB interface module (N2757A)
  Fully programmable with IEEE488.2 compliance
  Typical GPIB throughput of 20 measurements or twenty 2000-point records per second
- Optional printer kit
  The N2727A is a thermal printer kit, including printer power, parallel cable, power cable, and paper.

### D.2.15 General Characteristics

- Physical

  Size: 32.26 cm wide × 17.27 cm high × 31.75 cm deep (without handle)

  Weight: 6.35 kg (14 lbs) on 54621A/D, 54622A/D, 54624A; 6.82 kgs (15 lbs) on 54641A/D, 54642A/D
- Probe comp output

  54621A/D, 54642A/D, 54624A: Frequency ~1.2 kHz; Amplitude ~5 V

  54641A/D, 54642A/D: Frequency ~2 kHz; Amplitude ~5 V
- Trigger out

  54621A/D, 54622A/D, 54624A: 0 to 5 V with 50 source impedance; delay ~ 55 ns

  54641A/D, 54642A/D: 0 to 5 V with 50 source impedance; delay ~ 22 ns
- Printer power

  7.2 to 9.2 V, 1 A
- Kensington lock

  Connection on rear panel for security

### D.2.16 Power Requirements

- Line voltage range

  54621A/D, 54622A/D, 54624A: 100 – 240 VAC ±10%, CAT II, automatic selection

  54641A/D, 54642A/D: 100–240 VAC, 50/60 Hz, CAT II, automatic selection; 100–132 VAC, 440 Hz, CAT II, automatic selection
- Line frequency

  54621A/D, 54622A/D, 54624A: 47 to 440 Hz

  54641A/D, 54642A/D: 50/60 Hz, 100–240 VAC; 440 Hz, 100–132 VAC
- Power Usage

  54621A/D, 54622A/D, 54624A: 100 W max

  54641A/D, 54642A/D: 110 W max

### D.2.17 Environmental Characteristics

- Ambient temperature

  Operating –10°C to +55°C; nonoperating –51°C to +71°C
- Humidity

  Operating 95% RH at 40°C for 24 hr; nonoperating 90% RH at 65°C for 24 hr

- Altitude

  Operating to 4,570 m (15,000 ft); nonoperating to 15,244 m (50,000 ft)
- Vibration

  HP/Agilent class B1 and MIL-PRF-28800F; Class 3 random
- Shock

  HP/Agilent class B1 and MIL-PRF-28800F (operating 30 g, 1/2 sine, 11-ms duration, 3 shocks/axis along major axis. Total of 18 shocks)
- Pollution degree 2

  Normally only dry nonconductive pollution occurs. Occasionally a temporary conductivity caused by condensation must be expected.
- Indoor use only

  This instrument is rated for indoor use only

### D.2.18 Other

- Installation categories

  CAT I: Mains isolated

  CAT II: Line voltage in appliance and to wall outlet
- Regulatory information

  Safety:

  IEC 61010-1:1990+A1:1992+A2:1995/EN 61010-1:1994+A2:1995

  UL 3111

  CSA-C22.2 No. 1010.1:1992
- Supplementary information

  The product complies with the requirements of the Low Voltage Directive 73/23/EEC and the EMC Directive 89/336/EEC, and carries the CE-marking accordingly. The product was tested in a typical configuration with HP/Agilent test systems

## D.3 Operating Principle of HP 54510A Digitizing Oscilloscope

The HP 54510A Digitizing Oscilloscope is a very easy-to-use signal measuring equipment. The instrument can handle input signals from 0 to 100 MHz and has a very high resolution in that range. There are four input channels, which give it the capability to measure four signals simultaneously. The functional operating steps of the HP 54510A are as follows:

- Turn on the HP 54510 by pressing the power key at the *back* of the instrument.
- Connect the input signal to any of the four available channels: **channel 1, channel 2, channel 3 or channel 4**.
- Press **autoscale**, and observe the waveform on the display screen.
- To display the *peak-to-peak amplitude* of the signal, press **function key (blue key)** and press **Vp-p**.
- To display the *maximum amplitude* of the signal, press **function key (blue key)** and press **Vmax**.
- To display the *minimum amplitude* of the signal, press **function key (blue key)** and press **Vmin**.
- To display the *frequency* of the signal (if the signal is periodic), press **function key (blue key)** and press **freq**.
- To display the *time period* of the signal (if the signal is periodic), press **function key (blue key)** and press **period**.
- Press **save** to save the currently displayed waveform *to* memory, with a file name.
- Press **recall** to display any saved waveform *from* memory, by specifying the appropriate file name.
- Press **display** to adjust the format of the display, such as **gridlines** (**on** or **off**) and **dotted line**, or **full line** display.

## Reference

1. http://cp.literature.agilent.com/litweb/pdf/5968-8152EN.pdf

# Appendix E

## Texas Instruments DSPs and DSKs

### E.1 Introduction to Digital Signal Processors (DSPs)

A digital signal processor (DSP) is a type of microprocessor — one that is incredibly fast and powerful.[1] A DSP is unique because it processes data in real time, which makes it perfect for applications that cannot tolerate any delays. For example, did you ever talk on a cell phone where two people could not talk at once? You had to wait until the other person finished talking. If you both spoke simultaneously, the signal was cut — you did not hear the other person. With today's digital cell phones, which use DSP, you can talk normally. The DSP processors inside cell phones process sounds so rapidly you hear them as quickly as you can speak — in real time. Here are just some of the advantages of designing with DSPs over other microprocessors:

- Single-cycle multiply-accumulate operations
- Real-time performance, simulation, and emulation
- Flexibility
- Reliability
- Increased system performance
- Reduced system cost

The overall architecture of a typical DSP is shown in Figure E.1. The heart of the DSP is the Central Processing Unit (CPU), which is connected to the internal memory, external memory, and peripherals such as audio speakers and microphones.

Some material in this Appendix is reproduced with permission from Texas Instruments, Inc., Dallas, TX.

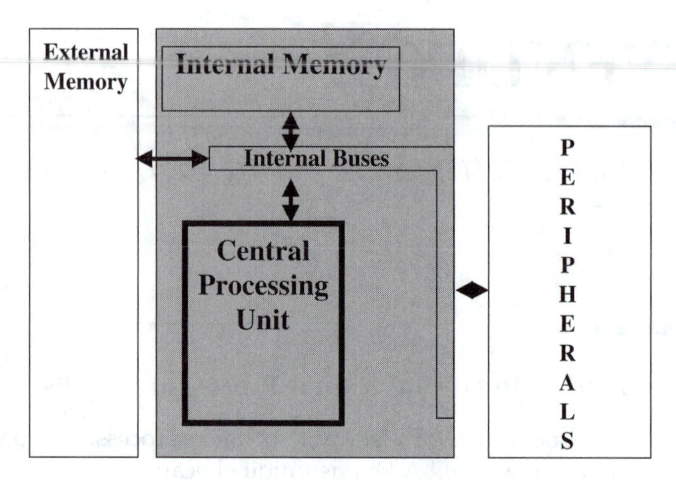


**FIGURE E.1**
Typical DSP block diagram.

### E.1.1 Alternative Solutions to Digital Signal Processors

Here is a look at some of the other alternatives available for digital signal processing and how they compare with DSPs.[1]

- The FPGA Alternative

  Field-programmable gate arrays (FPGAs) have the capability of being reconfigurable within a system, which can be a big advantage in applications that need multiple trial versions within development, offering reasonably fast time-to-market. They also offer greater raw performance per specific operation because of the resulting dedicated logic circuit. However, FPGAs are significantly more expensive and typically have much higher power dissipation than DSPs with similar functionality. As such, even when FPGAs are the chosen performance technology in designs such as wireless infrastructure, DSPs are typically used in conjunction with FPGAs to provide greater flexibility, better price/performance ratios and lower system power.
- The ASIC Alternative

  Application-specific Integrated Circuits (ICs) can be tailored to perform specific functions extremely well and can be made quite power efficient. However, because ASICS are not field programmable, their functionality cannot be iteratively changed or updated while in product development. As such, every new version of the product requires a redesign and trips through the foundry, an expensive proposition and an impediment to rapid time-to-market. Programmable DSPs, on the other hand, can be updated without changing the silicon; merely change the software program, greatly reducing development costs and availing aftermarket feature enhancements with mere code downloads. Consequently, when you see ASICs in real-time signal processing applications, they are typically employed as bus interfaces, glue logic, or functional accelerators for a programmable DSP-based system.

- The GPP Alternative

  In contrast to ASICs that are optimized for specific functions, general-purpose microprocessors are best suited for performing a broad array of tasks. However, for applications in which the end product must process answers in real time or must do so while powered by consumer batteries, GPPs' comparatively poor real-time performance and high power consumption all but rule them out. More and more, these processors are being seen as the dinosaurs of the industry — too encumbered with PC compatibility and desktop features to adapt to the changing real-time market place, as where world embraces tiny hand-held wireless-enabled products that require power dissipation measured in milliwatts — not the watts that these processors consume. Hence, DSPs are the programmable technology of choice. That trend is bound to continue as digital Internet appliances get smaller, faster, and more portable.

## E.2 Texas Instruments DSP Product Tree

The following list gives the TI DSP product line, starting with the latest product line.

- C6000

  TMS320C6000 High Performance DSPs deliver new levels of C-based performance and cost efficiency, with low power dissipation, for broadband networks and digitized imaging applications. Includes code compatible C62x and C64x fixed-point DSPs; C67x floating-point DSPs.

- C5000

  TMS320C5000 Power Efficient DSPs deliver an optimal combination of performance, peripheral options, small packaging, and the best power efficiency for personal and portable Internet and wireless communications. Includes code compatible C54x and C55x fixed-point DSPs.

- C2000

  TMS320C2000 Control Optimized DSPs deliver highest performance, greatest code efficiency, and optimal peripheral integration for the digital control revolution. Includes code compatible C24x and C28x fixed-point DSPs.

- OMAP

  OMAP Processors integrate the command and control functionality of ARM, coupled with low-power, real-time signal processing capabilities of a DSP. Optimized for mobile Internet devices and multimedia appliances.

- Other TMS320 DSPs

  All other TMS320 DSPs including C33x floating point DSPs.

## E.3 TMS320C6000™ Platform Overview Page

Raising the bar in performance and cost efficiency, the TMS320C6000™ DSP platform offers a broad portfolio of the industry's fastest DSPs, running at clock speeds up to 1 GHz. The platform consists of the TMS320C64x™ and TMS320C62x™ fixed-point generations as well as the TMS320C67x™ floating-point generation. Optimal for designers working on products such as targeted broadband infrastructure, performance audio, and imaging applications, the C6000 DSP platform's performance ranges from 1200 to 8000 MIPS for fixed point and 600 to 1800 MFLOPS for floating point.

### E.3.1 Code-Compatible Generations

With a broad portfolio of high-performance DSPs, the TMS320C6000 platform consists of three fully code-compatible device generations:

- **TMS320C64x**: The C64x fixed-point DSPs offer the industry's highest level of performance to address the demands of the digital age. At clock rates of up to 1 GHz, C64x DSPs can process information at rates up to 8000 MIPS, with costs as low as $19.95. In addition to a high clock rate, C64x DSPs can do more work each cycle with built-in extensions. These extensions include new instructions to accelerate performance in key application areas such as digital communications infrastructure and video and image processing.
- **TMS320C62x**: These first-generation fixed-point DSPs represent breakthrough technology that enables new equipments and energizes existing implementations for multichannel, multifunction applications, such as wireless base stations, remote access servers (RAS), digital subscriber loop (xDSL) systems, personalized home security systems, advanced imaging and biometrics, industrial scanners, precision instrumentation, and multichannel telephony systems.

- **TMS320C67x**: For designers of high-precision applications, C67x floating-point DSPs offer the speed, precision, power savings, and dynamic range to meet a wide variety of design needs. These dynamic DSPs are the ideal solution for demanding applications such as audio, medical imaging, instrumentation, and automotive.

### E.3.2 C Compiler

The C6000 DSP platform also offers a high-performance C engine with a compiler that leverages the architecture to sustain maximum performance while speeding design development time for high-performance applications. The C compiler/optimization tools balance code size and performance to meet the needs of an application and are available for download at no cost.

### E.3.3 C6000 Signal Processing Libraries and Peripheral Drivers

To enable designers to dramatically reduce code development time and enable faster time-to-market, platform-specific libraries have been developed, that can be downloaded at no cost. The Signal Processing and the Chip Support libraries contain a collection of high-level, optimized DSP function modules and help to achieve performance higher than standard ANSI C code.

## E.4 TMS320C6711 DSP Chip

The C6711 DSP chip is a floating-point processor that contains a CPU (central processing unit), internal memory, enhanced direct memory access (EDMA) controller, and on-chip peripherals.[2,3] These peripherals include a 32-bit external memory interface (EMIF), two multichannel buffered serial ports (McBSP), two 32-bit timers, a 16-bit host port interface (HPI), an interrupt selector, and a phase lock loop (PLL), along with hardware for *Boot Configurations* and *Power Down Logic*, as shown in Figure E.2.[5]

### E.4.1 Timing

The DSP chip must be able to establish communication links between the CPU (DSP core), the codecs, and memory. The two McBSPs, namely serial port 0 (SP0) and serial port 1 (SP1), are used to establish asynchronous links between the CPU and the on-board codec, and between the CPU and daughter card expansion, respectively. These McBSPs use *frame synchronization* to communicate with external devices.[5] Each McBSP has seven pins. Five of

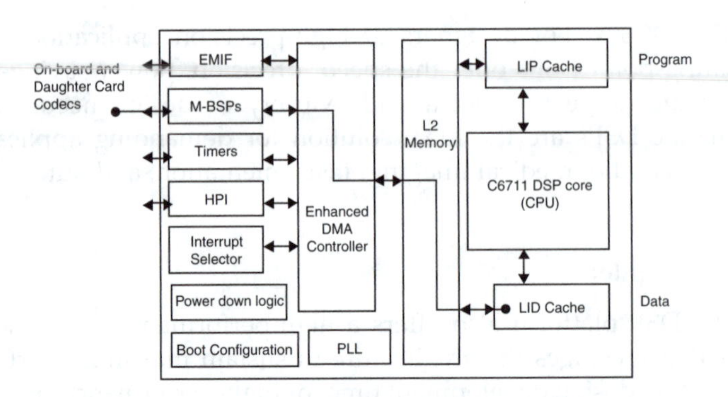


**FIGURE E.2**
TMS320C6711 DSP chip layout. (Courtesy of Texas Instruments Inc., Dallas, TX.)

them are used for timing, and the other two are connected to the data receive and data transmit pins on the on-board codec or daughter card. Also included in each McBSP is a 32-bit serial port control register (SPCR). This register is updated when the on-board codec (or daughter card) is ready to send data to or receive data from the CPU. The status of the SPCR will only be a concern to us when polling methods are implemented.

In the labs described in this book, we will be exploring two possible ways of establishing a *real-time* communication link between the CPU and the on-board codec. The idea of real-time communication is that we want a continuous stream of samples to be sent to the codec. In our case, we want samples to be sent at rate 8 kHz (one sample every .125ms). This is controlled by the codec, which will signal serial port 0 (SP0), every .125ms or, in the case of a daughter card, every $(1/24) * 10^{-3} = 0.0417$ ms.

- **Polling**

  The first method for establishing a real-time communication link between the CPU and the on-board codec is *polling*. When the on-board codec is ready to receive a sample from the CPU, it sets bit 17 of the SPCR in the McBSP on the DSP chip to true. Bit 17 of the SPCR is the CPU transmit ready (XRDY) bit, which the on-board codec uses to let the CPU know when it can transmit data2. In a polling application, the CPU continuously checks the status of the SPCR and transmits a data sample as soon as bit 17 of the SPCR is set to true. Upon transmission, the McBSP will reset bit 17 of the SPCR to false. The polling program will then wait until the on-board codec resets bit 17 to true before transmitting the next data sample. In this manner, a polling algorithm will maintain a constant stream of data flowing to the on-board codec.

  On the DSP hardware, polling is implemented mostly in software. The on-board codec will continuously set the transmit-ready bit of the SPCR, and the McBSP on the DSP chip will always reset it. However, it is up to the programmer to write a program that will continuously check the status of the SPCR. Fortunately, a program (support file) has already been written to manage this. The details of this program are explained in Chapter 7, where a polling example is implemented.

- **Interrupts**

  Interrupts are a way of handling asynchronous events on the DSP chip and can be generated either internally through software or externally by other components on the DSK. Handling (or servicing) of interrupts requires extra hardware that operates autonomously from the CPU. The C6711 chip is equipped with this hardware (timers, McBSP, etc.) and is the preferred way to time events on the DSP chip. By convention, a binary 0 is used for false and a binary 1 is used for true. The bits in the SPCR are labeled from the point-of-view of the DSP chip and not the peripheral with which it is communicating.

  In the labs detailed in Chapter 7, we use interrupts to establish a real-time communication link between the on-board codec and the CPU via SP0. When interrupts are used to establish the real-time link between the on-board codec and the CPU, the interrupt registers in SP0 are configured to handle interrupts. Then, when the codec sets the transmit-ready bit in the SPCR, the McBSP will generate an interrupt. When an interrupt occurs, the following events happen: the current program execution is halted; the current execution state is saved (in a CPU register); the program branches to and processes the interrupt; and upon completion of the interrupt, the execution state is restored and the program continues executing.

On the 'C6711 DSP chip, there are 32 possible interrupt sources, but only 12 may be assigned by the programmer, namely INT4 through INT15.[5] These 12 interrupts are prioritized by the "Interrupt Selector" (see Figure E.2). In these labs, we use the *SP0 transmit interrupt*, which is labeled by the interrupt acronym XINT0 in the TI literature.[5] Arbitrarily, we choose INT11 to handle this interrupt.[4] Using interrupts requires that each one be mapped to an interrupt service routine (ISR). This is done by a *vectors* file that, in our case, maps INT11 to the C-coded ISR c_int11(). In addition, INT11 must be selected to handle interrupts from XINT0, and the DSP chip must be set up to accept programmer assigned interrupts.[4] These tasks will be done by the C-coded function comm intr().

## E.5 TMS320C6711 Digital Signal Processing Starter Kit

The 'C6711 DSP Starter Kit, or DSK, provides system design engineers with an easy-to-use, cost-effective way to take their high-performance TMS320C6000 designs from concept to production. The new 'C6711 DSK not only provides an introduction to 'C6000 technology, but is also powerful enough to use for fast development of networking, communications, imaging, and other applications. Operating at 150 MHz, the 'C6711 delivers an impressive 1200 MIPs and 600 MFLOPs for only $22. The 'C6711 DSK replaces and is a superset of the 'C6211 DSK. The 'C6711 is binary code compatible with the 'C6211. C, assembly, and executable code written for the 'C6211 will run without modification on the 'C6711. The C6711 DSK board and its components are shown in Figure E.3.

### E.5.1 Hardware and Software Components of the DSK

Hardware

- 150 MHz 'C6711 DSP
- TI 16-bit A/D converter ('AD535)
- External memory
  - 16M bytes SDRAM
  - 128K bytes flash ROM
- LED's
- Daughter card expansion
- Power supply and parallel port cable

Software

- Code generation tools
  - (C compiler, assembler, and linker)
- Code composer debugger
  - (256 K program limitation)
- Example programs and utilities
  - Power-on self-test
  - Flash utility program
  - Board confidence test
  - Host access via DLL
  - Sample program(s)

The 'C6711 DSK comes with an array of DSK-specific software functionality (256 KB software image memory limited), including the highly efficient 'C6000 C compiler and assembly optimizer, code composer debugger, and DSK support software (flash utility, sample programs, and confidence tests). The daughter card interface socket provides a method for accessing most of the C6711 DSP for hardware extension.

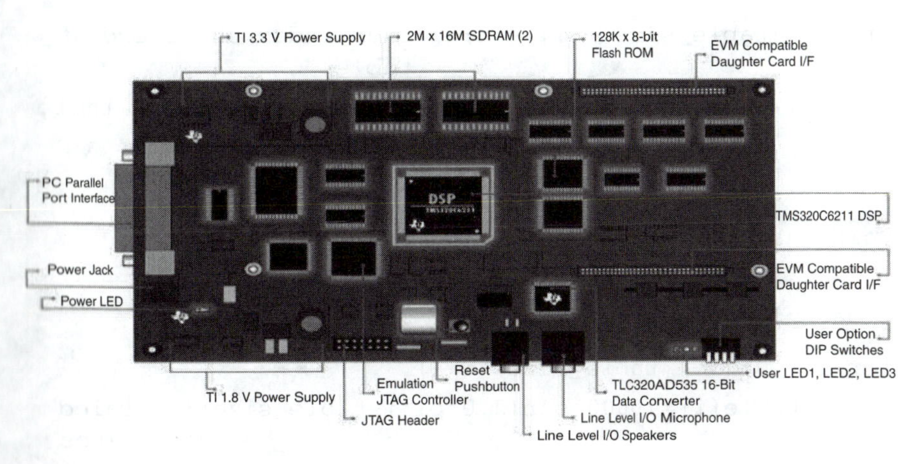


**FIGURE E.3**
Photograph of Texas Instruments TMS320C6711 DSK. (*Courtesy of Texas Instruments Inc., Dallas, TX.*)

## E.6 C Files for Practical Applications Using the 'C6711 DSK

This section lists C files for some practical DSP projects that are possible using the TMS320C6711 DSP Starter Kit. We can separate the applications discussed here into three broad categories: *signal generation* applications, *spectral analysis* applications using FFT, and *digital filtering* applications.

### E.6.1 Signal Generation Applications Using the 'C6711 DSK

- Sine wave generation using a look-up table

```
//Sinegen_table.c Generates a sinusoid for a look-up
table

#include <math.h>
#define table_size (short)10      //set table size
short sine_table[table_size];     //sine table array
short i;

interrupt void c_int11()  //interrupt service routine
{
output_sample(sine_table[i]);//output each sine value
```

```
if (i < table_size - 1) ++i; //incr index until end of
                                   table
 else i = 0;             //reinit index if end of table
return;                  //return from interrupt
}
void main()
{
float pi=3.14159;
for(i = 0; i < table_size; i++)
 sine_table[i]=10000*sin(2.0*pi*i/table_size);//scaled
                                                  values
i = 0;
comm_intr();        //init DSK, codec, McBSP
while(1);           //infinite loop
}
```

- Sine wave generation using MATLAB (Note that this application requires two files: Main C file **Sin1500MATL.c** and header file **sine1500.h**, both given below).

```
//Sin1500MATL.c Generates sine from table created with
MATLAB
#include "sin1500.h" //sin(1500) created with MATLAB
short i=0;
interrupt void c_int11()
{
output_sample(sin1500[i]); //output each sine value
if (i < 127) ++i; //incr index until end of table
 else i = 0;
return;    //return from interrupt
}
void main()
{
comm_intr(); //init DSK, codec, McBSP
while(1); //infinite loop
}
```



```
//sine1500.h header file
short sin1500[128]=
{0, 924, 707, -383, -1000, -383
383, 1000, 383, -707, -924,
0, 924, 707, -383, -1000, -383,
383, 1000, 383, -707, -924,
0, 924, 707, -383, -1000, -383,
383, 1000, 383, -707, -924,
0, 924, 707, -383, -1000, -383,
383, 1000, 383, -707, -924,
0, 924, 707, -383, -1000, -383,
383, 1000, 383, -707, -924,
0, 924, 707, -383, -1000, -383,
383, 1000, 383, -707, -924,
0, 924, 707, -383, -1000, -383,
383, 1000, 383, -707, -924,
0, 924, 707, -383, -1000, -383,
383, 1000, 383, -707, -924};
```

- Square wave generation using look-u

```
//Squarewave.c Generates a squar
table
#define table_size (int)0x100  /
int data_table[table_size]; //dat
int i;
interrupt void c_int11()   //inte
{
 output_sample(data_table[i]); //
 if (i < table_size) ++i; //if ta
 else i = 0;             //reinitia
 return; //return from interrupt
}
main()
{
 for(i=0; i<table_size/2; i++) //s
   data_table[i] = 0x7FFF; //with
 for(i=table_size/2; i<table_size;

   data_table[i] = -0x8000; //with
```



```
i = 0;
 comm_ir
 while (
}
```

- Ram

```
//Ramp.c
int outp
interrup
{
output_s
output
if (outp
output
return;
}
void ma
{
output
comm_in
while(1
}
```

- Pse
two
give

```
//Noise
#incluc
int fb;
shift_r
interru
{
int prr
```

```
if(sreg.bt.b0) //sequence{1,-1}based on bit b0
prnseq = -8000;          //scaled negative noise level
else
prnseq = 8000;           //scaled positive noise level
fb =(sreg.bt.b0)^(sreg.bt.b1); //XOR bits 0,1
fb ^=(sreg.bt.b11)^(sreg.bt.b13);//with bits 11,13
->fb sreg.regval<<=1;   //shift register 1 bit to left
sreg.bt.b0 = fb;                //close feedback path

output_sample(prnseq);     //output scaled sequence
return;                    //return from interrupt
}
void main()
{
```

```
//Noise_gen.h header file for pseudo-random noise
sequence
typedef struct BITVAL  //register bits to be packed as
                         integer
{
unsigned int b0:1, b1:1, b2:1, b3:1, b4:1, b5:1, b6:1;
unsigned int b7:1, b8:1, b9:1, b10:1, b11:1, b12:1,b13:1;
unsigned int dweebie:2;//Fills the 2 bit hole - bits 14-15
} bitval;
typedef union SHIFT_REG
{
unsigned int regval;
bitval bt;
} shift_reg;
```

### E.6.2 Spectral Analysis Applications Using the ‘C6711 DSK

- FFT calculator application (Note that this application requires two files: Main **C file FFT256c.c** file, which *calls* a callable FFT function file **FFT.c**, both given below).

```
//FFT256c.c FFT implementation calling a C-coded FFT
function

#include <math.h>
#define PTS 256          //# of points for FFT
#define PI 3.14159265358979
typedef struct {float real,imag;} COMPLEX;
void FFT(COMPLEX *Y, int n); //FFT prototype
float iobuffer[PTS];    //as input and output buffer
float x1[PTS];          //intermediate buffer
short i;                //general purpose index variable
short buffercount = 0; //number of new samples in
                          iobuffer
short flag = 0;         //set to 1 by ISR when iobuffer
                          full
COMPLEX w[PTS];         //twiddle constants stored in w
COMPLEX samples[PTS];      //primary working buffer

main()
{
for (i = 0 ; i<PTS ; i++)   //set up twiddle constants in w
{
 w[i].real = cos(2*PI*i/512.0);//Re component of twiddle
                                  constants
 w[i].imag =-sin(2*PI*i/512.0);//Im component of twiddle
                                  constants
}
comm_intr();                        //init DSK, codec, McBSP
while(1)                            //infinite loop
{
 while (flag == 0) ;        //wait until iobuffer is full
 flag = 0;             //reset flag
 for (i = 0 ; i < PTS ; i++)  //swap buffers
 {
  samples[i].real=iobuffer[i]; //buffer with new data
  iobuffer[i] = x1[i];     //processed frame to iobuffer
 }
 for (i = 0 ; i < PTS ; i++)
```

```
  samples[i].imag = 0.0;        //imag components = 0

  FFT(samples,PTS);        //call function FFT.c

  for (i = 0 ; i < PTS ; i++)  //compute magnitude
 {
  x1[i] = sqrt(samples[i].real*samples[i].real
 + samples[i].imag*samples[i].imag)/32;
 }
 x1[0] = 32000.0;      //negative spike(with AD535)for ref
}                            //end of infinite loop
}                            //end of main
interrupt void c_int11()   //ISR
{
output_sample((int)(iobuffer[buffercount])); //out from
                                                iobuffer
iobuffer[buffercount++]=(float)(input_sample());//input
                                              to iobuffer
if (buffercount >= PTS)        //if iobuffer full
{
buffercount = 0;               //reinit buffercount
flag = 1;                      //set flag
}
}
```

---

```
//FFT.c C callable FFT function in C

#define PTS 256                //# of points for FFT
typedef struct {float real,imag;} COMPLEX;
extern COMPLEX w[PTS];   //twiddle constants stored in w

void FFT(COMPLEX *Y, int N)   //input sample array, # of
                                points
{
COMPLEX temp1,temp2;      //temporary storage variables
int i,j,k;            //loop counter variables
int upper_leg, lower_leg;  //index of upper/lower
                             butterfly leg
```

```
int leg_diff;        //difference between upper/lower leg
int num_stages = 0; //number of FFT stages (iterations)
int index, step; //index/step through twiddle constant
i = 1; //log(base2) of N points= # of stages
do
{
 num_stages +=1;
 i = i*2;
}while (i!=N);
leg_diff = N/2; //difference between upper and lower legs
step = 512/N;   //step between values in twiddle.h
for (i = 0;i < num_stages; i++) //for N-point FFT
{
 index = 0;
 for (j = 0; j < leg_diff; j++)
 {
  for (upper_leg = j; upper_leg < N; upper_leg +=
                               (2*leg_diff))
  {
  lower_leg = upper_leg+leg_diff;
    temp1.real = (Y[upper_leg]).real +
                               (Y[lower_leg]).real;
  temp1.imag = (Y[upper_leg]).imag + (Y[lower_leg]).imag;
  temp2.real = (Y[upper_leg]).real - (Y[lower_leg]).real;
  temp2.imag = (Y[upper_leg]).imag - (Y[lower_leg]).imag;
  (Y[lower_leg]).real = temp2.real*(w[index]).real
                      -temp2.imag*(w[index]).imag;
  (Y[lower_leg]).imag = temp2.real*(w[index]).imag
                       +temp2.imag*(w[index]).real;
  (Y[upper_leg]).real = temp1.real;
  (Y[upper_leg]).imag = temp1.imag;
  }
 index += step;
 }
 leg_diff = leg_diff/2;
 step *= 2;
  }
```

```
  j = 0;
  for (i = 1; i < (N-1); i++)  //bit reversal for
                                resequencing data
  {
 k = N/2;
 while (k <= j)
  {
  j = j - k;
  k = k/2;
  }
 j = j + k;
 if (i<j)
  {
   temp1.real = (Y[j]).real;
   temp1.imag = (Y[j]).imag;
   (Y[j]).real = (Y[i]).real;
   (Y[j]).imag = (Y[i]).imag;
   (Y[i]).real = temp1.real;
   (Y[i]).imag = temp1.imag;
  }
 }
 return;
}
```

### E.6.3 Digital Filtering Applications Using the 'C6711 DSK

- Digital filtering with FIR filter

```
//Fir.c FIR filter. Include coefficient file with length N

#include "bs2700.cof"  //coefficient file BS @ 2700Hz
int yn = 0;                //initialize filter's output
short dly[N];      //delay samples

interrupt void c_int11()  //ISR
{
   short i;
```

```
   dly[0] = input_sample(); //new input @ beginning of
                                 buffer
   yn = 0;           //initialize filter's output
   for (i = 0; i< N; i++)
   yn += (h[i] * dly[i]); //y(n) += h(i)* x(n-i)
   for (i = N-1; i > 0; i—) //starting @ end of buffer
   dly[i] = dly[i-1];   //update delays with data move

   output_sample(yn >> 15); //scale output filter
   return;
}

void main()
 {
   comm_intr();        //init DSK, codec, McBSP
   while(1);          //infinite loop
 }
```

- Digital filtering with IIR filter

```
//IIR.c IIR filter using cascaded Direct Form II
//Coefficients a's and b's correspond to b's and a's from
MATLAB

#include "bs1750.cof"     //BS @ 1750 Hz coefficient file
short dly[stages][2] = {0};  //delay samples per stage

interrupt void c_int11()  //ISR
{
int i, input;
int un, yn;

input = input_sample();   //input to 1st stage
for (i = 0; i < stages; i++) //repeat for each stage
{
 un=input-((b[i][0]*dly[i][0])>>15) -
((b[i][1]*dly[i][1])>>15);

yn=((a[i][0]*un)>>15)+((a[i][1]*dly[i][0])>>15)+((a[i][2
]*dly[i][1])>>15);
```

```
dly[i][1] = dly[i][0];    //update delays
 dly[i][0] = un;      //update delays
 input = yn; //intermediate output->input to next stage
}
 output_sample(yn);     //output final result for time n
 return;                      //return from ISR
}
void main()
{
comm_intr();              //init DSK, codec, McBSP
while(1);                     //infinite loop
}
```

- Digital filtering with FIR *Notch* filter

```
//NOTCH2.C Two FIR notch filters to remove two sinusoidal
noise signals

#include "BS900.cof"      //BS @ 900 Hz coefficient file
#include "BS2700.cof"    //BS @ 2700 Hz coefficient file
short dly1[N]={0};     //delay samples for 1st filter
short dly2[N]={0};      //delay samples for 2nd filter
int y1out = 0, y2out = 0;  //init output of each filter
short out_type = 1;         //slider for output type

interrupt void c_int11()   //ISR
{
   short i;

   dly1[0] = input_sample();  //newest input @ top of
                                buffer
   y1out = 0;             //init output of 1st filter
   y2out = 0;             //init output of 2nd filter
   for (i = 0; i< N; i++)
  y1out += h900[i]*dly1[i]; //y1(n)+=h900(i)*x(n-i)

   dly2[0]=(y1out >>15);    //out of 1st filter->in 2nd
                              filter
   for (i = 0; i< N; i++)
```

```
    y2out += h2700[i]*dly2[i];//y2(n)+=h2700(i)*x(n-i)

   for (i = N-1; i > 0; i—)  //from bottom of buffer
   {
     dly1[i] = dly1[i-1];   //update samples of 1st buffer
   dly2[i] = dly2[i-1];   //update samples of 2nd buffer
  }

   if (out_type==1)        //if slider is in position 1
  output_sample(dly1[0]); //corrupted input(voice+sines)
   if (out_type==2)
  output_sample(y2out>>15); //output of 2nd filter (voice)
  return;                  //return from ISR
}

void main()
{
   comm_intr();           //init DSK, codec, McBSP
   while(1);           //infinite loop
}
```

- Digital filtering with *Adaptive* FIR filter

```
//AdaptIDFIR.c Adaptive FIR for system ID of an FIR (uses
C67 tools)
#include "bp55.cof" //fixed FIR filter coefficients
#include "noise_gen.h"   //support noise generation file
#define beta 1E-13 //rate of convergence
#define WLENGTH 60 //# of coefficients for adaptive FIR
float w[WLENGTH+1];       //buffer coeff for adaptive FIR
int dly_adapt[WLENGTH+1]; //buffer samples of adaptive FIR
int dly_fix[N+1];     //buffer samples of fixed FIR
short out_type = 1;       //output for adaptive/fixed FIR
int fb;              //feedback variable
shift_reg sreg;        //shift register

int prand(void)            //pseudo-random sequence {-1,1}
{
```

```
  int prnseq;
  if(sreg.bt.b0)
     prnseq = -8000;       //scaled negative noise level
  else
     prnseq = 8000;        //scaled positive noise level
  fb =(sreg.bt.b0)^(sreg.bt.b1);    //XOR bits 0,1
  fb^=(sreg.bt.b11)^(sreg.bt.b13); //with bits 11,13 -> fb
  sreg.regval<<=1;
  sreg.bt.b0=fb;              //close feedback path
  return prnseq;              //return noise sequence
}

interrupt void c_int11()         //ISR
{
int i;
int fir_out = 0;              //init output of fixed FIR
int adaptfir_out = 0;           //init output of adapt FIR
float E;                    //error=diff of fixed/adapt out

dly_fix[0] = prand();         //input noise to fixed FIR
dly_adapt[0]=dly_fix[0];      //as well as to adaptive FIR

for (i = N-1; i>= 0; i—)
 {
  fir_out +=(h[i]*dly_fix[i]); //fixed FIR filter output
  dly_fix[i+1] = dly_fix[i]; //update samples of fixed FIR
 }
  for (i = 0; i < WLENGTH; i++)
   adaptfir_out +=(w[i]*dly_adapt[i]);  //adaptive FIR
                                          filter output

  E = fir_out - adaptfir_out;      //error signal

  for (i = WLENGTH-1; i >= 0; i—)
   {
   w[i] = w[i]+(beta*E*dly_adapt[i]); //update weights of
                                        adaptive FIR
   dly_adapt[i+1] = dly_adapt[i];    //update samples of
                                       adaptive FIR

  }
```

```
if (out_type == 1)        //slider position for adapt FIR
  output_sample(adaptfir_out);  //output of adaptive FIR
filter
else if (out_type == 2)  //slider position for fixed FIR
  output_sample(fir_out); //output of fixed FIR filter
return;
}

void main()
{
int T=0, i=0;
for (i = 0; i < WLENGTH; i++)

  w[i] = 0.0;               //init coeff for adaptive FIR
  dly_adapt[i] = 0;         //init buffer for adaptive FIR
  }
  for (T = 0; T < N; T++)
   dly_fix[T] = 0;          //init buffer for fixed FIR

sreg.regval=0xFFFF;         //initial seed value
fb = 1;                     //initial feedback value
comm_intr();                    //init DSK, codec, McBSP
while (1);                      //infinite loop
}
```

## References

1. Texas Instruments homepage, http://www.ti.com.
2. *TMS320C6000 Programmer's Guide, SPRU198D,* Texas Instruments, Dallas, TX, 2000.
3. *CPU and Instruction Set Reference Guide, SPRU189F,* Texas Instruments, Dallas, TX, 2000.
4. Chassaing, R., *DSP Applications Using C and the TMS320C6x DSK,* Wiley, New York, 2002.
5. *TMS320C6000 Peripherals, SPRU190D*, Texas Instruments, Dallas, TX, 2001.
6. Kehtarnavaz, N. and Keramat, M., *DSP System Design Using the TMS320C6000,* Prentice Hall, Upper Saddle River, NJ, 2001.
7. *TMS320C6000 Code Composer Studio User's Guide, SPRU328B,* Texas Instruments, Dallas, TX, 2001.

# Appendix F

## *List of DSP Laboratory Equipment Manufacturers*

### F.1 Introduction to DSP Laboratory Equipment

Several pieces of test and measurement equipment, including oscilloscopes, signal generators, and signal analyzers have been used in the laboratories covered in this book. While we have restricted the equipment discussion to specific models that are available in our DSP lab at CSUS, there exist several alternative models that would be equally useful for these applications. This appendix discusses alternative equipment, either from the same manufacturer, or from other manufacturers of test and measurement equipment. The following sections list comparative information on the following types of equipment:

- Digitizing oscilloscopes
- Synthesized signal generators
- Dynamic signal analyzers
- Spectrum analyzers

### F.2. Digitizing Oscilloscopes

The oscilloscope model that is used in this book, for laboratory exercises, is the HP54501A 100 MHz digitizing oscilloscope. However, other models from HP-Agilent[1] are listed in Table F.1 and from Tektronix in Table 4.2.[2]

---

Some material in this Appendix is reproduced with permission from the following sources: (a) Agilent Technologies, Palo Alto, CA, (b) Tektronix Corporation, Beaverton, OR, and (c) National Instruments Corporation, Austin, TX).

**TABLE F.1**

Comparative HP-Agilent Oscilloscope Data

| Model | Bandwidth | Channels | Memory Depth | Sample Rate | Price (U.S. $) |
|---|---|---|---|---|---|
| 54621A | 60 MHz | 2 | 4 MB | 200 MSa/s | 3,110 |
| 54621D | 60 MHz | 2 + 16 | 4 MB | 200 MSa/s | 4,340 |
| 54622A | 100 MHz | 2 | 4 MB | 200 MSa/s | 3,712 |
| 54622D | 100 MHz | 2 + 16 | 4 MB | 200 MSa/s | 5,661 |
| 54624A | 100 MHz | 4 | 4 MB | 200 MSa/s | 5,516 |
| 54641A | 350 MHz | 2 | 8 MB | 2 GSa/s | 6,114 |
| 54641D | 350 MHz | 2 + 16 | 8 MB | 2 GSa/s | 8.695 |
| 54642A | 500 MHz | 2 | 8 MB | 2 GSa/s | 8,196 |
| 54642D | 500 MHz | 2 + 16 | 8 MB | 2 GSa/s | 11,465 |
| 54830B | 600 MHz | 2 | Up to 16 MB | 4 GSa/s | 13,619 |
| 54830D | 600 MHz | 2 + 16 | Up to 16 MB | 4 GSa/s | 17,257 |
| 54831B | 600 MHz | 4 | Up to 16 MB | 4 GSa/s | 18,839 |
| 54831D | 600 MHz | 4 + 16 | Up to 16 MB | 4 GSa/s | 22,444 |
| 54833A | 1 GHz | 2 | Up to 16 MB | 4 GSa/s | 14,493 |
| 54833D | 1 GHz | 2 + 16 | Up to 16 MB | 4 GSa/s | 18,231 |
| 54832B | 1 GHz | 4 | Up to 16 MB | 4 GSa/s | 22,717 |
| 54832D | 1 GHz | 4 + 16 | Up to 16 MB | 4 GSa/s | 26,678 |
| 54853A | 2.5 GHz | 4 | Up to 32 MB | 20 GSa/s | 39,466 |
| 54854A | 4 GHz | 4 | Up to 32 MB | 20 GSa/s | 49,998 |
| 54855A | 6 GHz | 4 | Up to 32 MB | 20 GSa/s | 68,570 |

## F.3 Synthesized Signal Generators

The signal generator model that is used in this book is the HP3324A synthesized sweep generator. However, models from HP-Agilent and other manufacturers are listed below.

### HP-Agilent Models

Table F.3 lists the relevant details of some of the newer HP-Agilent models of signal generators.[3]

### Tektronix Models

Table F.4 lists the relevant details of some of the Tektronix models of signal generators.[4]

### National Instruments (NI) Models

- NI PXI-5401 60 MHz Function Generator[5]
    - 16 MHz sine; 1 MHz square, triangle, and ramp
    - 9.31 MHz frequency resolution
    - 12-bit amplitude resolution

**TABLE F.2**

Comparative Tektronix Oscilloscope Data

| | TH5700 | TDS1000 | TDS2000 | TDS30008 | TDS50008 | TDS60008 | TDS70008 | CSA70008 | TDS870008 | CSA80008 |
|---|---|---|---|---|---|---|---|---|---|---|
| Channels | 2*** | 2 | 2, 4 | 2, 4 | 2, 4 | 4 | 4 | 4 | Up to 8 | Up to 8 |
| Bandwidth | 100–200 MHz | 60–100 MHz | 60–200 MHz | 100–500 MHz | 350 MHz–1 GHz | 6 GHz and 8 GHz | 500 MHz–7 GHz | 1.5–4 GHz | 2.5–70 GHz | 2.5–70 GHz |
| Rise time | 3.5–1.75 ns | 5.83–3.5 ns | 5.83–1.75 ns | 3.5 ns–700 ps | 1.15 ns–300 ps | 70 ps and 50 ps | 800–60 ps | 240–100 ps | 5 ps | 5 ps |
| Sample rate (max. real-time) | 500 M/S–1 GS/s | 1 GS/s | 2 GS/s | 1.25 GS/s–5 GS/s | 1 GS/s–5 GS/s | 20 GS/s on 4 | 5 GS/s–20 GS/s | 20 GS/s | 200 kS/s (sequential) | 200 kS/s (sequential) |
| Oscilloscope type | DSO | DSO | DSO | DPO up to 3600 wfms/s | DPO/DPX up to 100,000 wfms/s | DSO | DPO/DPX >400,000 wfms/s | DPO/DPX >400,000 wfms/s | Sampling | Sampling |
| Record length (maximum) | 2.5 kB | 2.5 kB | 2.5 kB | 10 kB | 16 MB | 32 MB | 64 MB | 64 MB | 4 kB | 4 kB |
| Trigger types | Edge, pulse, video, external, motor | Edge, video, pulse (glitch) | Edge, video, pulse (glitch) | Edge, video, logic pattern state, pulse (glitch, width, runt, slew rate), communication | Edge, video, logic pattern, state, setup/hold, pulse glitch, logic qualified, width, runt, time-out, transition, window, trigger delay, communication | Edge, logic (pattern, state, setup/hold), pulse (glitch, logic qualified, width, runt, time-out, transition), communication and serial pattern | Edge, logic (pattern, state, setup/hold), pulse (glitch, logic qualified, width, runt, time-out, transition), communication and serial pattern | All TDS70008 types | Edge, internal clock, clock recovery | Edge, internal clock, clock recovery |
| Connectivity | Basic | Advanced via optional TDS2CMA communications module | Advanced via optional TDS2CMA communications module | Extended | Extended | Extended | Extended | Extended | Extended | Extended |
| Application specific solutions | Power | Fast Fourier Transform (FFT) | Fast Fourier Transform (FFT) | Communication, video | Power, communication, video, jitter, disk drive, USB, ethernet, optical storage | Serial data, signal integrity, jitter, timing analysis | Serial data, signal integrity, jitter, timing analysis | Communication, jitter, disk drive, USB 2.0 | Serial data, communication including GBE, Fibre Channel, XAUI, SATA, PCI-Express | Serial data, communication including GBE, Fibre Channel, XAULT, SATA, PCI-Express |

**TABLE F.2 (continued)**

Comparative Tektronix Oscilloscope Data

| | TH5700 | TDS1000 | TDS2000 | TDS30008 | TDS50008 | TDS60008 | TDS70008 | CSA70008 | TDS870008 | CSA80008 |
|---|---|---|---|---|---|---|---|---|---|---|
| Waveform math and analysis** | Basic | Basic, plus FFT standard | Basic, plus FFT standard | Basic, plus FFT standard | Extended | Extended | Extended | Extended | Extended | Extended |
| Other features | Handheld, battery power | External trigger input, autoset menu, probe check wizard, automatic measurements, multi-language user interface | External trigger input, autoset menu, probe check wizard, automatic measurements, multi-language user interface | Portable (7 lbs/3.2 kg) battery power | CD-RW drive, 512 MB RAM, keyboard | Computer and Datacom mask testing, serial pattern trigger, both with clock recovery, front panel USB 2.0 port | High resolution (XGA) display, graphical user interface, triggered and untriggered roll modes | Built-in optical reference receiver, clock recovery, mask testing | Optical and electrical sampling modules with integrated clock recovery | Optical and electrical sampling modules with integrated clock recovery |
| Applications | Power harmonic measurements, installation, maintenance and repair | Service and repair, education and training, manufacturing test and quality control, design and debug | Service and repair, education and training, manufacturing test and quality control, design and debug | Telecommunication mask testing and manufacturing, digital design/troubleshooting, video design/service, power supply design | Digital design and debug, power measurements, video design, DVD analysis | Validation/characterization/compliance testing of high-speed digital designs, jitter analysis, serial data analysis | Validation/characterization/compliance testing of high-speed digital designs, jitter analysis, disk drive measurements power electronics, communication mask testing | Design development, optical and electrical compliance testing, signal integrity, margin verification, jitter and timing analysis | Device characterization and semiconductor testing, cross-talk characterization, TDR and TDT | High-speed tele and data communications, signal analysis and compliance testing |

**TABLE F.3**

Comparative HP-Agilent Signal Generator Data

| Model | Signal Type | Maximum Frequency | Number of Outputs | Configuration Price (U.S. $) |
|---|---|---|---|---|
| 33120A | Arbitrary and standard waveforms including sine, square, triangle, ramp, and noise | 15 MHz for sine and square waves | Single channel | 2,153 |
| 8114A | Pulses at voltages up to 100 V and currents up to 2 A | 15 MHz | Single channel | 12,308 |
| 33220A | Arbitrary and standard waveforms including sine, square, pulse, triangle, ramp, and noise | 20 MHz for sine and square waves | Single channel | 1,975 |
| 81101A | Pulses | 50 MHz (transition times from 5 ns to 200 ms) | Single channel | 6,653 |
| 33250A | Arbitrary and standard waveforms including sine, square, pulses, triangle, ramp, and noise | 80 MHz for sine and square waves, 50 MHz for pulses, 25 MHz for arbitrary waveforms | Single channel | 4,682 |
| 81104A | Pulses and RZ/NRZ patterns including PRBS sequences | 80 MHz (transition times from 3 ns to 200 ms) | One or two channels | 13,056 |
| 81110A | Pulses and RZ/NRZ patterns including PRBS sequences | 165 MHz or 330 MHz (transition times from 800 ps to 200 ms; depending on output module) | One or two channels | 18,267 |
| 81130A | Pulses and RZ/NRZ patterns including PRBS sequences | 400 MHz or 600 MHz (transition times from 500 ps to 1.6 ns; depending on output module) | One or two channels | 21,530 |
| 81133A | Pulses and RZ/NRZ patterns including PRBS sequences | 3.35 GHz (transition times from 60 to 120 ps (typical)) | Single channel | 47,570 |
| 81134A | Pulses and RZ/NRZ patterns including PRBS sequences | 3.35 GHz (transition times from 60 to 120 ps (typical)) | Two channels | 62,103 |

- Frequency sweeps and hopping with 4 triggering modes
- 16 KB memory for arbitrary waveform generation
- NI-FGEN driver software optimized for use with LabVIEW and LabWindows/CVI

- NI PXI-5404 100 MHz Function Generator[6]
  - 9 kHz to 100 MHz sine wave generation
  - DC to 100 MHz clock generation
  - 1.07 μHz frequency resolution
  - ±0.2 dB flatness across sine wave passband (9 kHz to 100 MHz)
  - 12-bit amplitude resolution
- NI PXI-5411 40 MS/s Arbitrary Waveform Generator[7]
  - 1 channel
  - 16 MHz sine wave generation; SYNC (TTL) outputs
  - 12-bit resolution
  - 2 or 8 million sample waveform memory; 4 triggering modes
  - Waveform linking and looping; waveform and frequency hopping
  - IVI-compliant NI-FGEN driver
- NI PXI-5421 100 MS/s, 16-Bit Arbitrary Waveform Generator[8]
  - 12 VP-p into 50 Ω load
  - Up to 400 MS/s effective sampling rate with interpolation
  - 91 dBc close-in SFDR and –62 dBc THD at 10 MHz
  - –148 dBm/Hz average noise density
  - 8, 32, or 256 MB of onboard memory
  - Optional 16-bit LVDS digital pattern output

## F.4 Dynamic Signal Analyzers

The signal analyzer model that is used in this book is the HP35665A Dynamic Signal Analyzer. However, other models, from HP-Agilent and other manufacturers are listed below.

**HP-Agilent Model[9]**

- 35670A 2 or 4 channel FFT Dynamic Signal Analyzer, DC-102.4 kHz
  - 102.4 kHz at 1 channel, 51.2 kHz at 2 channel, 25.6 kHz at 4 channel
  - 100, 200, 400, 800, and 1600 lines of resolution
  - 90 dB dynamic range, 130 dB in swept-sine mode

**TABLE F.4**

Comparative Tektronix Signal Generator Data

| Product | Channels (max.) | Sample Rate (max.) | Memory Depth (max.) | Vertical Resolution (bits) | Output Amplitude (max.)* | Marker Outputs (max.) | Parallel Digital Outputs (max.) | Integrated Editors | Built-In Applications | Complementary Products |
|---|---|---|---|---|---|---|---|---|---|---|
| AWG710B | 1 | 4.2 GS/s | 64.8 M | 8 | 2 | 2/ch | — | G, E, S | DD, NPL, JG | TDS/CSA7000B, TDS6000 series oscilloscopes, TLA logic analyzers |
| AWG615 | 1 | 2.7 GS/s | 64.8 M | 8 | 2 | 2/ch | — | G, E, S | DD, NPL, JG | |
| AWG500 series | 2 | 1.0 GS/s | 4 M | 10 | 2 | 2/ch | 10 | G, E, S | | |
| AWG400 series | 3 | 200 MS/s | 16 M | 16 | 5 | 2/ch | 48 | G, E, S | NPL, JG, DM | TDS5000B, TDS3000B series oscilloscopes; TLA logic analyzers |
| AWG2021 series | 2 | 250 MS/s | 256 k | 12 | 5 | 2/ch | 24 | G, E, S | — | |
| AWG2005 series | 4 | 20 MS/s | 64 k | 12 | 10 | 1/ch | 24 | G, E, S | — | |
| AFG300 series | 2 | 16 MS/s | 16 k | 12 | 10 | 1 (sync) | — | T | — | TDS3000B, TDS2000, TDS1000 series oscilloscopes; TLA logic analyzers |

*Note:* Integrated editors: G = graphical, E = equation; S = sequence; T = text.

Built-in features: DD = disk drive; NPL = network physical layer; JG = jitter generation; DM = digital modulation.

* VP-p into 50 Ohm.

- Source: random, burst random, periodic chirp, burst chirp, pink noise, sine, arbitrary waveform
- Measurements: linear, cross, power spectrum, power spectral density, frequency response, coherence, THD, harmonic power, time waveform, auto-correlation, cross-correlation, histogram, PDF, CDF
- Octave analysis with triggered waterfall display
- Tachometer input and order tracking with orbit diagram
- Built-in 3.5-inch floppy disk drive

**National Instrument (NI) Model**

- NI PCI-4551, NI 4552 Dynamic signal Analyzers[10,11]
  - 2 or 4 analog input
  - 90 dB dynamic range, 16-bit resolution
  - –20 to 60 dB gains
  - 204.8 kS/s maximum sampling rate
  - Analog triggering, 32-digital I/O lines
  - Embedded 32-bit digital signal processor
  - 100, 200, 400, 800, 1600 lines FFT resolution
  - Windowing — Hanning, Blackman-Harris, Kaiser, exponential, uniform, user-defined
  - Operating Systems
    - Windows 2000/NT/XP/Me/9x
  - Recommended software
    - LabView
    - LabWindows/CVI
    - Real-Time Octave Analysis for NI-DSA
  - Other compatible software
    - Visual Basic
    - CC++
  - Driver software (included)
    - NI-DSA

## F.5 Spectrum Analyzers

The spectrum analyzer model that was used in the laboratory exercises for this book is the HP8590L Spectrum Analyzer. However, other models are listed below.

## HP-Agilent Models

Table F.5 lists the relevant details of some of the newer HP-Agilent models of spectrum analyzers.[12]

**TABLE F.5**

Comparative HP-Agilent Spectrum Analyzer Data

| Product | Frequency Range | RF/MW Performance (Form Factor) | Key Attributes | Available Measurement Personalities and Software |
|---|---|---|---|---|
| PSA series spectrum analyzers | 3 Hz–50 GHz (external mixing to 325 GHz) | Highest performance (bench top) | Highest dynamic range, accuracy, flexibility and connectivity, wireless format digital demodulation | Noise figure, phase noise, W-CDMA, HSDPA, GSM/EDGE, 1xEV-DO, 1xEV-DV, cdma2000, cdmaOne, NADC, PDC, TD-SCDMA |
| 8560EC series spectrum analyzers | 3 Hz–50 GHz (external mixing to 325 GHz) | Highest performance (portable) | High dynamic range in a rugged portable package | Phase noise, spurious response, digital radio |
| ESA series express analyzers | 9 kHz–26.5 GHz | Mid-performance to basic performance (portable) | Fast delivery, easy selection | Noise figure, phase noise, fault location, GSM, GPRS, EDGE, cdmaOne, EVM analysis, cable TV |
| ESA-E series spectrum analyzers | 100 Hz–26.5 GHz (external mixing to 325 GHz) | Mid-performance (portable) | Scalable price/performance, great dynamic range and accuracy, wireless format digital demodulation | Noise figure, phase noise, Bluetooth, fault location, CATV/broadcast TV, GSM, GPRS, cdmaOne, EVM analysis |
| ESA-L series spectrum analyzers | 9 kHz–26.5 GHz | Basic performance (portable) | Affordable speed and accuracy for general-purpose spectrum analysis | Cable TV |
| E4406A vector signal analyzer | 7 MGz–4 GHz | Mid-performance (bench top) | Fast, wireless format digital demodulation, baseband I&Q measurements | W-CDMA, HSDPA, GSM/EDGE, 1xEV-DO, 1xEV-DV, cdma2000, cdmaOne, NADC, PDC, iDEN |

**TABLE F.5 (continued)**

Comparative HP-Agilent Spectrum Analyzer Data

| Product | Frequency Range | RF/MW Performance (Form Factor) | Key Attributes | Available Measurement Personalities and Software |
|---|---|---|---|---|
| 89600 series vector signal analyzers | DC to 6 GHz (to 50 GHz with spectrum analyzers) | Mid-performance (PC based-software with VXI hardware) | Flexible, in-depth digital modulation analysis; >36 MHz bandwidth; ADS link for software simulation | 802.11a/b, Bluetooth, TETRA, W-CDMA, GSM/EDGE, 1xEV-DO, cdma2000, TD-SCDMA, PHP, digital video |
| 89400 series vector signal analyzers | DC to 2.65 GHz | High performance (bench top) | Flexible, in-depth digital modulation analysis; 8 MHz bandwidth | Bluetooth, TETRA, GSM, cdmaOne, NADC, PDC, PHP, digital video |
| 89601A vector signal analysis software | 3 Hz–50 GHz (combined with spectrum analyzers) | Mid-performance to high performance (PC based software with spectrum analyzer or oscilloscope hardware) | Flexible in-depth digital modulation analysis; up to 6 GHz bandwidth; ADS link for s/w simulation | 802.11a/b/g, Bluetooth, TETRA, W-CDMA, GSM/EDGE, 1xEV-DO, cdma2000, TD-SCDMA, PHP, digital video |
| Related spectrum analyzer products | n/a | n/a | n/a | n/a |
| 4395A/4396B combination network/spectrum/impedance analyzers | 10 Hz–500 MHz and 100 kHz–1.8 GHz | Mid-performance spectrum analyzer (bench top) | General-purpose spectrum, vector network, and impedance analysis in one box | Time gating spectrum analysis |

## Tektronix Models

Table F.6 lists the relevant details of some of Tektronix models of spectrum analyzers.[13]

**TABLE F.6**

Comparative Tektronix Spectrunm Analyzer Data

| RSA Model | Frequency Range | Memory Depth | Modulation Analysis | Real-Time Capture Bandwidth | Triggering Modes |
|---|---|---|---|---|---|
| 2203A | 10 MHz–3 GHz, DC to 3 GHz (option 05) | 2 MB | AM, FM (ASK, FSK), PM | 10 MHz | IF level |
| 2208A | 10 MHz–8 GHz, DC to 8 GHz (option 05) | 2 MB | AM, FM (ASK, FSK), PM | 10 MHz | IF level |
| 3303A | DC to 3 GHz | 64 MB, 256 MB (option 02) | AM, FM (ASK, FSK), PM; general purpose digital modulation analysis (option 21) | 15 MHz | IF level; frequency mask trigger and power (span BW) (option 02) |
| 3308A | DC to 8 GHz | 64 MB, 256 MB (option 02) | AM, FM (ASK, FSK), PM; general purpose digital modulation analysis (option 21) | 15 MHz | IF level; frequency mask trigger and power (span BW) (option 02) |

## References


1. http://cp.literature.agilent.com/litweb/pdf/5968-8152EN.pdf.
2. http://www.tek.com/Measurement/cgi-bin/framed.pl?Document=/Measurement/scopes/comparison.html?wt=257&link=/Measurement/scopes/comparison.html&FrameSet=oscilloscopes.
3. http://cp.literature.aglent.com/litweb/pdf/5968-8807EN.pdf.
4. http://www.tek.com/Measurement/cgi-bin/framed.pl?Document=/Measurement/signal_sources/home.html?wt=257&link=/Measurement/signal_sources/home.html&FrameSet=signal_sources.
5. http://sine.ni.com/apps/we/nioc.vp?cid=11941&lang=US.
6. http://sine.ni.com/apps/we/nioc.vp?cid=12024&lang=US.
7. http://sine.ni.com/apps/we/nioc.vp?cid=1884&lang=US.
8. http://sine.ni.com/apps/we/nioc.vp?cid=12472&lang=US.
9. http://cp.literature.agilent.com/litweb/pdf/5966-3064E.pdf.
10. http://sine.ni.com/apps/we/nioc.vp?cid=1487&lang=US.
11. http://sine.ni.com/apps/we/nioc.vp?cid=1488&lang=US.
12. http://cp.literature.agilent.com/litweb/pdf/5962-7275E.pdf.
13. http://www.tek.com/site/ps/0,,37-17118-SPECS_EN,00.html.

TABLE E.5

[illegible]

| RSA Model | Frequency Range | Maximum Distance | Modulation Method | Bandwidth | Signaling Rate |
|---|---|---|---|---|---|
| [illegible] | [illegible] | 2 MB | [illegible] | [illegible] | [illegible] |
| [illegible] | [illegible] | [illegible] | [illegible] | [illegible] | [illegible] |
| [illegible] | [illegible] | [illegible] | [illegible] | [illegible] | [illegible] |
| [illegible] | [illegible] | [illegible] | [illegible] | [illegible] | [illegible] |

Endnotes

[illegible]

# Index

## C

## E



## F

## G



## H

# I

## N



## O



## P

## Q



## R



## S

## T

## U



## V



## W



## Y



## Z